LAST CHANCE TO LAUNCH

HOW LEADERS RESCUE ERP, CRM, AND DIGITAL-TRANSFORMATION PROJECTS FROM THE EDGE OF FAILURE

DUSTIN DOMERESE

LAST CHANCE TO LAUNCH:
HOW LEADERS RESCUE ERP, CRM, AND DIGITAL-TRANSFORMATION
PROJECTS FROM THE EDGE OF FAILURE

Copyright © Dustin Domerese (2025)

ISBN Paperback: 979-8-89576-167-0
ISBN Hardback: 979-8-89576-168-7

Published by:

🖋 AUTHORS on MISSION

Dedication

To God, whose words and grace have illuminated every step of the journey;

To my pastor, whose wisdom and spiritual leadership have strengthened my walk;

To my gracious parents, whose faith helped me believe that all things are possible with God;

To my beautiful wife, whose unwavering love and support made this pursuit possible;

To my children, who have graciously shared precious time with this purpose;

And to so many colleagues and clients, many of whom live within these pages, whose partnership, challenges, chaos, and brilliance have shaped both my career and many of the lessons captured in this book.

Contents

A Posture of Humility

"Your project isn't just behind schedule. After two years and triple the original budget, you have nothing in production. Worse still, if you complete this implementation with the current software, the complexity could paralyze your operations department."

The conference room fell silent as I delivered the news.

The executives exchanged worried glances, their faces reflecting a mix of frustration and resignation. They had invested millions in this new system implementation; they promised shareholders a digital transformation that would increase efficiency, reduce overhead, and automate processes, but now they were forced to face the prospect of joining the ranks of other failed IT projects that litter the corporate landscape.

This scene plays out hundreds of times across organizations and repeats itself daily in boardrooms around the world. I've witnessed many large IT projects fail to deliver promised value: it would seem that for every successful digital transformation, two others are failing to reach their goals, burning billions in investment in their wake.

Why? The answer may surprise you.

While I have seen both successes and failures while guiding hundreds of technology implementations, a crucial lesson has emerged: technical issues are seldom the root cause of failure. They exist, of course, but they aren't what usually throws a project off course.

The real culprits?

1. First, there is always a human element to digital transformations. The technology is supposed to fix the problem

in areas like poor data quality, inconsistent processes, and minimal performance tracking, but systems cannot improve what hasn't been first defined and resolved manually by humans. Because humans are involved, the inherently imperfect and rational logic does not always prevail within the organizations.

2. Second, human inefficiency arises when there are misaligned expectations. Companies often believe that software will make their lives easier, but poor definition of goals and concrete requirements can overcomplicate projects, causing them to fail in their promises, or, even worse, sometimes erase what makes the business unique.

3. Third, resistance to change due to organizational dynamics can bring emotion and ego into what should be logical and systematized. Power struggles, competing priorities, and the expectation that consultants will bring transformation in their backpacks can stall progress. Breaking down data and department silos and focusing on internal capabilities are keys to success.

4. More importantly, complexity management is inherent in such projects and must be owned and embraced by all members of the team. Unclear timelines, undefined roles, and unrealistic budgets result in confusion. Adding overly complicated solutions that are usually poorly documented results in a system that doesn't satisfy the needs of a business, costs more than it should, and takes longer than it needs to.

In my work with Dynamic Consultants Group, I've had the privilege of seeing spectacular successes and enlightening failures. Projects literally on the brink of disaster were eased back from the precipice and transformed into examples of how to do digital transformation the right way. What was fundamentally different? The answer is simple, but the execution is hard. Move the first priority focus away from technical solutions and toward managing people, processes, and overall organization objectives.

Consider this real-world example of how even an apparently straightforward automation can turn into a nightmare once its basic principles are forgotten:

Case Study

The warehouse floor was in complete disarray.

Beeping scanners, frustrated voices, and the far-off clatter of metal reverberated around the huge space like a storm brewing.

Workers dashed between towering shelves, their faces contorted with confusion and rising panic. The state-of-the-art $2 million ERP system that was supposed to be the future of their operations had now created chaos.

One employee, with fifteen years at the company under his belt, slammed his barcode scanner onto a nearby workstation. "This is *insane!*" His voice rang through the aisles, thick with exasperation. "The system says we've got *500* units of the J-47 connectors in Bin A-13. *Look at it!*" He gestured wildly at the empty bin. "Meanwhile, I've got three bins *overflowing* with parts that don't even show up in the system!"

Uneasy glances were exchanged by colleagues around him, some having long given up in scanning as futility set in. The attempt to reconcile the digital with the physical inventory kept on getting worse. Order processing once seemed simple but now had ground to a halt with these inventory nightmares.

Ripple effects pushed outward.

On the factory floor, two entire assembly lines stood still. Silent, lifeless. Workers, usually a blur of activity, leaned against machinery or paced in frustration. Without the right components, they couldn't build anything. Every minute wasted was money bleeding out from the organization.

In the shipping department, pallets of half-completed outbound orders stacked higher by the hour, untouched. Customers were waiting. Trucks sat idle at the loading docks while drivers checked their watches or paced outside their trucks complaining to one another.

In customer service, it seemed like the phones never stopped ringing.

"What do you mean you *can't find* my order? *I placed it two weeks ago*!" a furious client barked over the line.

"I—I understand, sir. We're experiencing—" the representative fumbled. "We're working on a resolution—"

The caller had already hung up.

Upstairs, the plant manager's office had become the project war room. Sarah Chester, the executive who had spearheaded this transformation, sat at the head of the conference table, fingers pressed into her temples. Around her, a shouting match raged.

"This *isn't working*!" the manufacturing lead yelled. "We need to go back to the paper logs *right now*!"

"The software is *fine*!" the IT director snapped. "This is a *user error* issue. Your warehouse workers can't adapt to the technology."

"We're *drowning* down there!" The warehouse supervisor cut him off. "My guys don't even know what they're scanning! We need *training*!"

Sarah's heart was racing. The weight of a multimillion-dollar failure seemed to press upon her chest. It had worked *perfectly* in the test environment. The vendor had *guaranteed* this was the best solution on the market. So how could this have gone wrong so quickly?

It wasn't only the technology that had a problem.

For *years*, different shifts had developed their own ways of doing things. Unofficial part names, undocumented workarounds, bin labeling that changed from one aisle to the next. The warehouse functioned on tribal knowledge, passed down through whispered instructions and instinct, not structured data and process. It wasn't that the system had broken the company; it had exposed the deep cracks that had always been there.

Now, those cracks had widened into what seemed like chasms.

They automated their own dysfunction. Instead of fixing broken processes, they had amplified them. Work orders were now being triggered based on *wrong* inventory data, causing unnecessary orders for surplus parts while essential components ran dry.

To Sarah, this realization landed like a gut punch.

This mess wouldn't get fixed with another software patch or an IT directive forcing system processes. It would take something far harder: tearing down the informal, undocumented practices that had kept things running, standardizing them, and rebuilding trust—not in the technology, but in the processes themselves.

I have seen this very lesson being worked out on a global scale with too many businesses to remember. Be it inventory management, ERP and CRM systems, or financial management software and reporting, the same truth emerged:

Technology doesn't fix broken organizations. It reveals *just how broken they already were.*

This story demonstrates a number of key points which we will discuss in greater detail during this book:

1. First, technology amplifies existing problems. The new ERP didn't create chaos; it just uncovered it. What had previously been manageable mislabeling and inventory discrepancies

became a full-scale crisis when automated tracking forced reliance on the flawed data and processes.

2. Second, tools are only as good as the foundation under them. This warehouse had run for years on undocumented tribal knowledge, informal workarounds, and inconsistent labeling. Without standardizing these first, the system had no reliable framework on which to be built. No consultant or vendor in the world can solve these issues unless an organization adopts a posture of humility and a willingness to change.

3. Third, rushing to implementation based on arbitrary timelines and unrealistic budgets leads to prolonged disruption. The company rolled out the new system before the workers were ready and before the processes had been refined. What was to have been a leap toward efficiency turned out to mean months of operational paralysis.

In this book, we'll examine how to avoid these pitfalls and build transformations that truly transform—not just automate dysfunction. The stakes are too high for trial and error, so it's important to understand how to avoid these pitfalls and instead build transformations that *make a positive difference in the trajectory of an organization*.

Whether it's deploying a new ERP system, implementing CRM platforms, executing on-premise to cloud migrations, or adopting advanced supply chain and warehouse management solutions, the basic hurdles encountered remain surprisingly similar. Technical challenges inevitably arise. However, these challenges rarely stem from inherent software defects. Instead, they typically originate from human factors: insufficient testing processes, inadequate implementation planning, misaligned business goals, unclear expectations, and gaps in technical specifications. When systems fail, it's often because these foundational elements weren't properly addressed.

This book presents a radical proposition: because technical challenges exist, and because projects fail when not managed properly, successful project delivery does not rely solely on technology. It hinges on how you manage people through strong leadership, cultivate aligned stakeholders, and implement effective processes that drive performance measurement, operational excellence, and strategic automation—all while maintaining your organization's competitive advantages. Whether starting fresh or rescuing a failing implementation, the principles in this book remain the same.

The principles we will review are not theoretical. Rather, they have been forged in the furnaces of real-world scenarios. My personal experience has centered within the Microsoft and SAP ecosystem, where I have trained hundreds of partners and implemented solutions for clients from small businesses to Fortune 500 companies; these insights prove to be universal.

In the chapters ahead, we'll learn:

- How to lay proper groundwork before project initiation
- Strategies for aligning stakeholders for long-term success
- The critical role of empowerment and goal setting
- Crafting your Project Bill of Rights to open the organization's communication
- How to establish ground rules for all parties involved in the implementation
- The revolutionary 90-day implementation philosophy
- Methods for sustaining momentum through implementation
- Frameworks for ensuring user adoption and training, and assisting team members in learning new skills
- How to successfully navigate inevitable challenges that will occur

This guide focuses on the human and process aspects of digital transformation and therefore goes beyond what another technical system implementation guide(s) might offer. This book offers actionable steps required for managing the interplay among people, processes, and technology.

The stakes couldn't be higher. With the rapid evolution of today's business sector, ERP, CRM, and other digital transformation initiatives have become a fundamental requirement for survival. The companies that master these principles will thrive. Those that don't risk contributing to the growing graveyard of failed digital initiatives.

Before embarking on an implementation, it is crucial to define clearly the underlying issues at hand. Just as any system's strength depends on its underlying support, organizational preparedness must be addressed thoroughly before turning to technological fixes.

Through the proven methodologies outlined in the next chapters, you can dramatically improve your odds of success. The principles in this book have helped hundreds of organizations navigate complex implementations successfully. They can help you too.

CHAPTER 1

Understanding the Core Problem(s)

"There never was a good knife made of bad steel."
—Benjamin Franklin, Poor Richards Almanack

B y 2026, organizations worldwide are expected to waste over *$2 trillion* on failed digital transformations (Whitmore, 2023). This figure reflects a broader trend where ambition often outpaces execution, leaving behind abandoned server rooms as the silent monuments to their unrealized potential. Much like attempting to build a house on quicksand, the effort to implement enterprise software on shaky organizational foundations is destined to fail. Without solid project groundwork, even the most well-planned transformations will crumble, underscoring the critical need for a balanced approach that aligns strategy with the capabilities needed for success.

In order to prevent this outcome, enterprise system implementations need first to expose organizational weaknesses rather than expecting the solution itself to resolve them. It's important to understand what the problem is before attempting an implementation. A system can only be as strong as the foundation it's built upon, making it essential to address organizational readiness before diving into technical solutions.

Let's start by looking at the current landscape of organizational failures. From there, we'll delve deeply into the four core problems that consistently thwart digital transformation projects, costing organizations millions in wasted investment and lost productivity.

Project Failure Statistics Across Industries

First, it is important to understand the sheer volume of organizations that attempt digital transformations across a wide range of industries. Despite large investments in technological advancement by many organizations, the rates of their success remain painfully low.

The Standish Group's analysis of 50,000 global technology projects reveals that 66% end in partial or total failure (Standish Group International, 2020). More concerning, McKinsey's research indicates that 17% of large IT projects become so problematic that they threaten organizational survival (Bloch et al., 2012). The Boston Consulting Group found an equally troubling trend, saying 70% of digital transformation efforts fall short of their intended targets (BCG, 2020).

These failures come at a great cost. According to Harvard Business Review, IT projects run, on average, 27% over budget; one in six turns out to be a "black swan"—in other words, catastrophically bad—that runs 200% over budget. Not only that, but 70% of them are over schedule. Kmart's $1.2B failed IT modernization is a case in point; it directly contributed to its bankruptcy (Flyvbjerg et al., 2023).

These sobering statistics reinforce one important fact: digital transformation success requires a lot more than just the proper selection of technologies. In a bid to avoid becoming another cautionary tale, organizations must understand and address four core problems that continuously derail these initiatives. Let's understand each in detail.

Core Problem #1: The Human Element

It is no secret that digital transformation in modern days is rapidly increasing. Automated systems benefit everyone in one way or another. Business leaders use dashboards to make informed decisions based on data. IT teams automate routine maintenance tasks. Sales teams instantly access customer information through mobile apps. Even the owner of your local coffee shop uses point-of-sale systems to track inventory and analyze peak hours. But nice as it would be to say that technology can solve your problems with a touch of a button: You can't automate what people haven't understood and mastered. Even in the new world of large language models, (LLM) you only need to do some simple, human-generated prompts to understand that the better the human knowledge behind the prompt or request, the better the system will be at providing accurate information.

Indeed, many organizations chase technology implementation for perceived shortcuts to operational excellence through visions of systems that will automatically grant them complete data visibility, real-time insights, standardized processes, and consistency in execution. In fact, this idealistic vision will most assuredly hit operational realities head-on.

Rather than creating excellence, system implementations typically uncover these chronic examples of organizational immaturity:

- Data exists in scattered spreadsheets or databases with inconsistent formats.
  - *Example:* A sales team maintains customer information across dozens of spreadsheets, each formatted differently. Divisions or departments have their own version of the data, making cross-departmental reporting challenging if not impossible.
- Basic processes vary wildly between teams and individuals.
  - *Example:* Three customer service representatives handle refund requests using three different methods. These three representatives send the request to the accounting department, which processes the refunds in three different ways. This equates to nine possible ways to process a refund.
- Staff resist following structured workflows.
  - *Example:* Account managers and senior leaders bypass the internal ticketing system by sending their client's requests through email directly to an engineer. The moment this occurs you now have data silos that exist and inaccurate

reporting across the various processes in client engagement and customer service.

- Performance metrics don't exist or aren't trusted.
  - *Example:* Each regional office calculates sales metrics like win rate or conversion rate using different formulas. Then, even with an aligned way to perform the calculations, if every sales representative handles probability to close or recurring revenue calculations slightly differently when building quotes, the result will be major differences in metrics across departments.
- Different departments use different terminology for the same concepts.
  - *Example:* Marketing calls it a campaign, sales calls it a promotion, and finance calls it a program. These terms are not always industry standard, sometimes overlap with language used by another department for a different concept, and in general create a false sense of maturity that does not actually exist in the organization.

Consider it from this perspective.

Case Study

Phones fell silent in London first.

At BlackBerry's operations center, technicians watched as alerts began flooding their monitors. Within hours, the digital silence spread across Europe, then cascaded through the Middle East and into Africa. By the time it reached the Americas, millions of BlackBerry devices worldwide had fallen dark. No emails, no messages, no web access (Arthur & Baxter-Reynolds, 2011).

In the BlackBerry command center, the mood was tense. Engineers were leaning over screens, their faces illuminated by an array of blinking dashboards that revealed an ever-spreading network outage.

"It's just a switch failure," one senior technician assured his colleagues. "Our backup systems will handle it."

But they didn't.

Hours turned into days. CEOs around the globe weren't getting life-and-death emails, and it was costing organizations millions, if not billions, of dollars in lost productivity.

Hospitals were losing touch with on-call staff. Government officials were cut off from critical communications. The company that built its reputation on reliability—whose devices were so addictively dependable they were nicknamed CrackBerries—was in the throes of a crisis that would shake it to its core (Hayes, 2021).

For context, in the early 2000s, BlackBerry revolutionized how business was conducted. Their mobile devices, with their iconic physical keyboards and secure messaging capabilities, became indispensable tools for professionals worldwide. If you worked in business between 2000-2010, you likely either owned a BlackBerry or knew someone who did. These devices were so prevalent that BlackBerry served over 70 million subscribers globally (Seth, 2024).

Behind closed doors, however, a more insidious story emerged. Years of success and accelerating growth had led to complacency. Warning signals about infrastructure vulnerability had been waved off. "We're BlackBerry," was the common refrain. "Our systems don't fail."

Departments acted in silos, with vital information about the health of their systems trapped inside territorial boundaries. The culture, once driven by innovation and dependability, had calcified into rigid hierarchies, arrogance of risk, and change resistance.

The immediate cause was deceptively simple: a core switch in BlackBerry's internet servers had failed. The backup system, designed to seamlessly fail over, remained stubbornly silent, but the real failure wasn't due to hardware; it was the fault of leaders who had underestimated risks, departments that didn't communicate, and a culture that had lost its hunger for excellence (Arthur & Baxter-Reynolds, 2011).

Technological debt, sloppy testing processes, a lack of accountability, and leaders who did not take the time to understand the interworking of their own businesses created a cascade of failure points that were too complex to deconstruct quickly when the failure struck.

Then came the four days of darkness that underscored the many consequences of years of poor internal organizational performance.

When service resumed, BlackBerry had lost more than mere time; they had lost customer trust and market value. Their brand reputation lay in tatters. The company that once ruled mobile communications had finally shown its Achilles' heel—not in its technology, but in its people and processes.

The BlackBerry outage serves as a dramatic reminder: digital failures seldom happen directly because of technology per se. They are the consequence of human factors, complacency, breakdowns in communication, and rigidity in culture. Technology has a way of bringing to light the root issues that arise inside of an organization.

Today, in another conference room in another company, it is starting all over again. The question is not whether these companies will have their BlackBerry moment, but when—and if—they will realize what's really happening: a human problem masquerading as a technical one. Systems can't automate chaos into order.

The truth is, every organization has its BlackBerry moment. What separates winners from losers, however, is that they understand those moments for what they are, possess the wisdom and fortitude to handle them, and have processes in place to treat the underlying conditions before they translate into cataclysmic events.

Note:

The human element, by nature, will be a factor in all of the core problems presented below.

However, it stands as its own core problem for three key reasons:

1. Organizational humility fundamentally shapes willingness to change across all organizational levels but starts from the top down.
2. Human behavior patterns, ego, and personal priorities or goals determine system and process adoption success or failure.
3. Individual and team dynamics impact every aspect of implementation, but those dynamics must be governed by strong leadership and clearly defined core processes.

Core Problem #2: Misaligned Expectations

"If we take the CRM system out of the box, it will transform our sales process overnight!"

A system cannot be seen as a shortcut to processes. The idea that this mythical CRM system will come with a defined way to market, sell, and drive revenue for your organization seems ludicrous, but there are many organizations that, even if they will never say it out loud, will act according to that misguided belief during vendor selection or implementation. This approach reflects a critical misunderstanding of a project's expectation: simplicity to the end users or business stakeholders means choosing not to see the inherent complexity that exists in their own organization. Unfortunately, these metaphorical rose-colored glasses create a gap between vision and reality, and this results in devastating disappointment when reality cannot meet these inflated expectations, which often leads to abandoning projects or expanding the scope of work over budget.

The expectation gap manifests in several ways:

- Leadership envisions instant efficiency gains, but implementation often creates temporary inefficiency as users adapt.
  - *Example:* A marketing team expects instant mastery of a new email automation platform without accounting for the learning curve for simply using the system, understanding the terminology, and developing organization-specific campaign structure and journey mapping.
- Requirements exist mainly in people's heads rather than documented processes.
  - *Example:* IT implements new or changed features based on their own assumed requirements, or on requirements given only by senior executives, but never formally documented and never validated by the frontline workers or end users.
- RFPs focus on technical features while ignoring critical business process changes that must be made, as well as organizational change management that needs to occur.
  - *Example:* A warehouse or manufacturer focuses their RFP requirements on technical specifications like the need for scanners while ignoring critical workflow dependence on people to execute the very things they are asking for, like lot tracking and quality control.
- Promised outcomes (30% increase in sales) lack concrete implementation paths to achieve the stated goals or have no stated goal that is measurable and attainable.
  - *Example:* Sales leadership promises territory growth based on a new CRM implementation without defined adoption strategies, understanding of how it will increase velocity in the lead generation process, or having a defined sales follow-up process.

Standard software threatens to eliminate unique competitive advantages.

Example: A logistics company loses its rapid-response capability by forcing its fleet drivers and sub-contractors into an industry standard software workflow which requires many additional steps, latent internet connectivity, or additional data point capture in order to perform the same functions.

Look at it from this angle. If a global distribution company's leadership selects and implements an off-the-shelf warehouse management system, they probably expect it to provide a general capability of their cross-docking process, which gives them a market advantage. Instead, the standardized software is likely unable to accommodate any customization of their exact workflow processes, forcing them to either give up their competitive advantage or extensively customize outside of the system, neither of which aligns with their initial expectations.

Every great business has its secrets, its special sauce for competitive advantage. Many organizations have not done the market research and client surveying required to know what their advantages are. Standard software inherently drives process standardization. Organizations must protect the things that make them unique while embracing standardization where it is beneficial. This fine-tuning cannot be accomplished with software alone and must, from the outset, be a strategic objective of a technology initiative that takes into account and preserves an organization's uniqueness and advantages.

Core Problem #3: Organizational Dynamics

Behind every digital transformation failure, there's a story of organizational change resistance. As humans, whenever change appears to threaten our hand-carved roles, established ways of working, or even our own perceived value to the organization, we instinctively protect the things that matter most to us. The head of a department who has worked for many years to develop processes for their team will not let go easily. It is natural for an expert in a current system to be concerned about becoming a beginner again. It is natural for a manager who has gained influence due to specialized knowledge to be concerned about losing this hierarchical advantage.

Sometimes these human responses manifest as active opposition and sometimes as passive non-compliance, but they create far more decisive outcomes for the project than any technical software limitation ever could. It's rarely malice, but simply self-preservation. When people feel their expertise, autonomy, or security might be threatened, they will naturally focus their attention on looking after their own personal interests, the status quo of their department, and their own career paths.

Three critical challenges dominate:

1. Power struggles create departmental silos.
   o *Example:* Marketing wants to keep their lead scoring system, sales controls prospect qualification, and customer service won't share data about service to sales conversations because each is protecting their perceived domain of expertise.
2. Resource allocation becomes a battleground.
   o *Example:* The most experienced process expert in a manufacturing company cannot attend key system design sessions because production requires their constant attention, leaving junior staff to make critical decisions and the key team member to feel left out and isolated in the process.
3. Competing priorities derail progress.
   o *Example:* Department heads commit resources to the project but repeatedly pull them back for allegedly urgent operational needs, creating constant implementation delays.
4. Over-reliance on external expertise.
   o *Example:* Organizations delegate key decisions to consultants who lack deep understanding of internal processes, resulting in solutions that don't fit actual

business needs. The system then falls short of expectations from those that were not involved in the design or goal-setting process. Delegation becomes dysfunction.

These organizational dynamics, rooted in human nature and self-preservation instincts, can't simply be bulldozed through or ignored. Accomplishing all that, however, requires artful navigation and reflective leadership. The skills around understanding potential resistance and converting it to collaborative energy must be unearthed, developed, and empowered within a healthy organization.

Core Problem #4: Complexity Management

ERP and other digital transformations are inherently complex undertakings. They involve multiple departments, distinct workstreams, various vendors, diverse technology solutions, and numerous consultants. The sheer scale and scope of these projects mean that it only takes about two weeks for a project to become mired in details unless the complexity is managed with a disciplined and rigorous process. Without an iron fist on these elements, the project risks stalling and failing to meet its objectives before it can even lift off the launch pad.

Effective project management requires juggling many moving pieces while keeping direction crystal clear. However, for most organizations, large-scale transformation inherently means juggling unwieldy complexity that clouds direction and causes confusion, delays, and missed objectives.

Complexity challenges usually manifest as:

1. Unclear decision-making frameworks
   o *Example:* A retail chain's regional offices, corporate IT, and store operations all claim authority over system configuration, creating gridlock on crucial decisions.
2. Role confusion between stakeholders

- Example: Consultants, internal IT, and business units all believe the other owns the solution design documentation, leading to paralyzed progress and a misaligned expectation of time involvement and costs.

3. Overcomplicated solutions to simple problems
   - *Example*: An organization replaces its entire ERP system when all that was needed were better reporting tools, dramatically increasing project complexity and risk.
4. Hidden costs and unrealistic implementation timelines
   - *Example*: Project budget and timeline planning overlook critical costs and constraints like process redesign, training time, and lost productivity during transition periods.
5. Technical discussions overshadow business needs
   - *Example*: IT teams focus on the system's future capabilities, security, hardware, software licensing, performance, and features while losing sight of the basic business problems they're trying to solve in the short-term.

Digital transformation is complex on many levels, not just technical but also human, organizational, and strategic. To be successful, a framework is needed to wrangle these various dimensions into manageable pieces while keeping the big picture in sight. The chapters that follow discuss how to approach each of these essential issues in depth, but perhaps the most basic is laying a good foundation before the work starts. Complex transformations can be achieved when each of these challenges is broken down into strategic and manageable steps. They will never be achieved by accident but require intentionality on the part of the organization embarking on such a journey. Organizations with good but poorly defined intentions will be part of the all-too-common failure statistics.

It is fitting to imagine again the silent server rooms, once-blinking lights, once-whirring fans, now darkened and silenced following the many abandoned ERP and CRM systems. These rooms, once humming with new ambition and the promise of digital change, become monuments to misalignment, indecision, and transformation attempts gone off the rails. Every dark server rack is a testament to the cost of failure and lessons learned the hard way. What I offer in the following chapters is my attempt to remember what happens when project complexity outpaces project clarity, and why deliberate foundations, the humans involved, and organizational disciplines matter most.

Moving Forward

Understanding these four core problems reveals a crucial insight: successful digital transformation requires seamlessly integrating people, processes, and technology through communication and strategic vision.

Next up, we will uncover the key approaches for achieving sustainable alignment within organizations. You'll learn how to align visionary goals with tactical execution, build decision-making systems that promote clarity, and structure teams to encourage collaboration.

Organizations that master these elements transform implementation challenges into the operational excellence they may have been missing for years.

Reflection Questions
1. What unique competitive advantages does your organization possess that must be preserved through any system implementation? Has your organization performed the diligence required to confirm what it believes is its market advantage?
2. How do current organizational power dynamics between departments impact information sharing and decision-making? Write down names of those involved in the power dynamics and make a communication plan for each individual.
3. What processes in your organization still rely heavily on individual knowledge rather than documented procedures? Are you relying on leadership to confirm current processes, or have you interviewed frontline workers to confirm the processes are being followed?

4. Are your documented standard operating processes followed by everyone all the time? If not, make a list of all exceptions and identify if it is a people, process, or technology problem that prohibits the process from being followed all the time.
5. What specific measurable metrics would define success for your implementation beyond technical go-live? Make sure this metric is tied to increased revenue, reduced expenses, or added organizational efficiency and time savings in some way.

CHAPTER 2

The Human(s) in the Loop

"Technology is not the solution to organizational problems—it's a magnifying glass that makes them more visible." —Peter Drucker

D igital transformation will either succeed or fail based on one critical factor: the humans involved.

While technical challenges inevitably happen in any implementation, they rarely determine project outcomes. Instead, success will rely on how well organizations understand and address the human elements of change.

This principle remains crucial today, particularly in the age of AI and machine learning. Just as a human must master a task before teaching it to others, organizations must excel at manual processes before attempting to automate them. AI systems, despite their sophistication, learn from human expertise and patterns. If the underlying human processes are flawed or inconsistent, automation will only magnify these issues.

Think about Denver International Airport's baggage system in the 1990s. This can be seen as a cautionary tale in attempting automation before mastering human processes.

In the 1990s, Denver International Airport attempted to implement a fully automated baggage handling system. This ambitious project aimed to revolutionize how baggage was processed, but it failed spectacularly due to a lack of understanding and optimization of the human workflows it was meant to replace. The automated system was plagued with technical glitches, misrouted luggage, and mechanical failures. According to a 2005 article in The New York Times, the system's issues caused a $2 billion loss and delayed the airport's opening by 16 months. This failure has since become a case study of the dangers of relying on technology without first ensuring that the underlying human processes are efficient and well-understood.

Understanding these elements is essential because technology implementations don't just change systems—they change how people work, communicate, and deliver value. When organizations fail to address these human factors, even perfectly engineered solutions will fail to deliver intended results. Technology might be improving efficiency, but we must remember that its true purpose is to *help* people.

Why do technology projects matter to the humans or clients that they affect? It all comes down to a few core reasons:

- They provide better tools to get work done more efficiently.
- They free up time for workers to do more important tasks.
- They improve decision-making with better data.
- They help people collaborate and communicate more easily.

Technology must fully understand and align with human objectives. These elements form the foundation of an organization. We must not lose sight of these root benefits as they are the reason we engage with technology initiatives. Because of the close human connection, successful technology projects must consider both the people and the systems. Organizations that only think about the technical side, without considering how it affects people, often end up with well-built systems that don't work in practice.

Personal Posture: Are the Humans Really Ready?

It is also important to remember the psychology of personal versus organizational goals. One fundamental human psychological tendency is that we prioritize personal goals over organizational strategy. It's human nature, and it's also human nature to deny it. So, understanding and even articulating this dynamic publicly is important for project success.

Consider these psychological drivers:

- Self-preservation: People naturally protect their established ways of working because they represent proven paths to personal success that could be diminished by change.
- Status protection: Expertise in current systems provides status and influence that people fear losing when things change.
- Risk aversion: Known inefficient processes feel safer than unknown efficient ones.
- Achievement attribution: People tend to attribute their success to personal methods rather than organizational systems.

These factors manifest in behaviors like:

- Sales teams maintaining shadow spreadsheets that track their individual way of engaging
- Managers resisting standardization to maintain decision-making autonomy
- Leaders defending their direct reports against changes in process to be seen as heroes fighting for them against a new system
- Department heads protecting specialized processes that give them leverage
- Teams creating workarounds that serve their goals but fragment organizational data

Consider this real-world example of how misaligned expectations and poor human factor considerations can derail even the most technically sound implementations:

Case Study

The executive boardroom buzzed with tension as the implementation leader stood before a wall of screens displaying adoption metrics that told a story no one wanted to hear.

"Six months post-launch, and we're seeing only 20% consistent usage of the new CRM system," she reported, voice steady despite the growing unease in the room. "More concerning, our sales pipeline visibility has actually decreased by 40% compared to our pre-implementation baseline."

The CEO's face went dark. They had invested $3.8 million in this CRM implementation, expecting business-altering complete pipeline visibility, automated forecasting, and improved customer insights. Instead, they were drowning in conflicting data and shadow systems. Worse yet, many key sales team members were refusing to use the new system.

"That's impossible," the IT Director interjected. "The system is working perfectly. All technical metrics are green. The integration tests—"

"Show me the numbers for the Northeast region again," interrupted the VP of Sales, his voice carrying a note of challenge. The screen shifted to reveal an adoption rate of just 12%.

"These numbers can't be right," he continued. "That region consistently exceeds quota."

Another screen revealed what had been discovered by the system telemetry: over 300 individual spreadsheets managed by the sales

---

teams had been modified in the last 48 hours. In other words, there was a parallel system, fully independent of the CRM.

The room erupted in accusations:

"The sales team never bought in—"

"The training wasn't adequate—"

"The system is too rigid—"

The lead CRM consultant stretched out her hand. The room fell silent. "Yesterday, I sat with the Northeast team," she said, ensuring she had everyone's ears. "They showed me their 'real' system. It's not just spreadsheets. They've built an intricate web of relationship tracking honed over years—client preferences, personal details, informal agreements—none of which fit neatly into our standardized CRM fields."

A heavy silence fell over the room as the implications now sank in.

"Our top performers aren't resisting change out of stubbornness," she continued. "They're protecting something they believe is essential to their success. And here's the kicker—they're right."

Another slide appeared. "The teams with the lowest CRM adoption are consistently our highest performers. They're not failing to adapt to the system; the system is failing to adapt to what makes them successful."

The CEO leaned forward. "So what's the real problem here?"

The problem came into focus fast. "We applied a technical solution to a human problem. We never addressed the deep-seated mentality of our sales force. To our best team members, standardizing in the CRM means giving up what makes them great. We never asked them what they needed."

A new data set appeared on the screen. "Look at this. In the Western region, which has 78% adoption, the sales director did things differently. Prior to rolling out the CRM, he and his team spent three months analyzing how they managed their relationships. They configured the system to augment, rather than replace, their current practices. And what happened? The numbers for both adoption and sales performance rose.

One region had forced a system upon its processes; the other had absorbed it.

"The system isn't the problem," they concluded. "Our approach is. We can't force mindset change through technical implementation. We need to communicate with our people and reshape our approach."

Recognition settled over the room. Millions had been invested in technology, but the most crucial investment, securing buy-in from the people using it, had been overlooked in most regions.

The CEO finally spoke. "What do you recommend?"

She offered a three-month transformation plan, which would start not with technical training but rather with understanding—truly learning how top performers built and maintained customer relationships. The CRM would be reconfigured to support those practices rather than disrupt them.

A reminder, if nothing else, of one of the most basic facts of digital transformation: the best technology cannot overcome resistance to change, and the most elegant system will fail without true buy-in from those it is built to serve.

This may appear as a failure in a technology or engineering project. In reality, though, the main issues pertain to lack of alignment in organizational posture. This story teaches us that technical excellence alone is not enough to make the adoption successful, because many times high performers develop subtle workflows that standard systems don't capture. Forcing standardization without an understanding of current practice leads to resistance and sometimes unanticipated outcomes. True success arises when systems adapt and augment, not replace, proven methods. The best implementations start with regarding how people work versus just how the systems operate.

Digital transformation projects succeed not because of technology but because of people. True digital transformation happens when companies focus on people's needs first, develop long-term capabilities, and create a culture that supports change while keeping core values.

Organizational Posture: Is Your Organization Really Ready?

How do we know when an organization is ready for change, having a clear purpose, the right mindset, and good habits that can be repeated? We cannot forget to look at the personnel and the psychology behind why projects succeed or fail.

Organizational posture refers to how a company positions itself in its readiness for change, its cultural alignment, and its ability to adapt to new ways of working (Imtiaz, 2023). It ranges from leadership commitment and employee mindset to operational flexibility and change management capabilities. Consider this question as you assess your organizational posture for change: Are you rigid or resilient and ready? This one factor fundamentally informs how well organizations implement new systems and processes. Pre-implementation therefore is critical for an organization to assess its genuine preparedness for change.

Consider the following example of how organizational posture can make or break a digital transformation initiative:

Fictional Example

The executive boardroom of the manufacturing company hummed with fluorescent tension. PowerPoint slides clicked by, each laden with promises about their new ERP system's capabilities.

"And with these automated workflows," concluded the representative from the vendor, "in six months, you'll have a 40% efficiency gain."

The CEO nodded to that, already envisioning himself announcing this digital transformation to the board. "How soon can we start?"

Across town, the leadership team of another manufacturing company sat in a very different meeting. Their conference room walls were plastered with process maps, pain points scribbled in red marker, and notes from shop floor interviews.

"Before we even look at vendors," their Operations Director said, standing before the assembled group, "we need to answer some hard questions." She pointed to a particularly dense cluster of sticky notes. "Our third shift has completely different inventory procedures than first shift. Our quality control process varies by supervisor. And half our tribal knowledge isn't documented anywhere."

"But the software will standardize all that, right?" someone cut in.

"That's just the kind of thinking that will sink us," she said. "Software doesn't fix broken processes; it amplifies them."

Six months later, the contrast was stark.

In the first company, chaos reigned. The warehouse manager stormed out of yet another "optimization" meeting, his voice echoing down the hallway: "How did we get this far along and no one asked us how we actually track custom orders! This system is trying to force us into a box that will kill our business!"

Meanwhile, HR discovered their decades-old hiring workflows couldn't map to the new system.

The finance team had retreated to Excel spreadsheets, maintaining shadow systems that would never match their existing reports. Implementation deadlines slipped week after week.

"Just mandate that everyone use it," the CEO demanded in a particularly heated meeting.

"They'll quit first," the HR director warned. "We're already losing people."

The second company had its own set of problems, but the reaction was a whole different story. When the production team voiced concerns about new scanning procedures, they weren't dismissed. Instead, supervisors from all shifts huddled in a war room, mapping out their current processes step by step.

"Show me exactly how you handle rush orders," their Operations Director asked, marker in hand. The resulting workflow became the basis for the system configuration, not the other way around.

A year and a half later, the first company's chief executive officer had to attend the most dismal board meeting—millions over budget, core modules yet to go live, the CFO had left, and HR reported employee turnover was at an all-time high. Their initiative had resulted in only one major transformation: to the bottom line, and in the wrong direction.

The second company wasn't perfect, but their system was live, stable, and, most importantly, being used. The difference wasn't technical expertise or even budget. It was organizational posture. One company had tried to force their people into a system. The other had built a system around their people and processes. They showed a posture for the right standard processes being the center of the implementation.

What is your current organizational stance in view of change? What is your readiness for adaptation? And what is the overall alignment of leadership around the principle that the frontline decides success even before the first line of code has been written?

Understanding and addressing these foundational elements of organizational posture will create the bedrock for a successful transformation.

Project Posture: Purpose and Values

Even with proper positioning, organizations must still articulate clear purpose and values to guide their journey. Organizations need to spell out much more than operational requirements. A successful project begins with defining and communicating its purpose and values. It is at this step that you clearly communicate to your employees the overall purpose and values that will govern the implementation. Simultaneously, it communicates to the executive stakeholders what your business is all about: which market to serve and on what path the business will continue to realize its vision.

When you articulate the purpose and values, it will create a deep resonance between the organizational objectives and individual aspirations of your people. This personal and organizational alignment is the rocket fuel for a company to move from good to great in its digital transformation journey. These goals must be definable, measurable, attainable, and rooted in transformational business objectives rather than board room buzzwords that fail to deliver real results. Here are some real-world examples of what some of the goals should be for a digital transformation project:

1. Reduce clicks and improve efficiency
    - A medical device company reduced its sales order entry from 32 clicks to 7, letting sales reps spend 40% more time with customers.

- A manufacturing firm automated quality checks, shifting work with QA staff from data entry to complex problem-solving.
- An insurance company cut claims processing steps to only the ones absolutely required and shaved 60% off the time it took to open a claim. This enabled agents to handle a much higher volume of cases. Increasing agent throughput decreased labor costs and allowed for quicker claim resolution, higher NPS, higher retention rates, and increased profits.

2. Increase market share and revenue
   - A distribution company automated off the shelf orders, helping them expand market share by 20% while keeping the same staffing levels and without increasing delivery costs.
   - A software company streamlined their quote process from 3 days to 15 minutes, increasing sales team capacity by 65%.
   - A retailer unified their customer data through loyalty programs, enabling personalized marketing that boosted revenue 30%.

3. Boost customer satisfaction
   - A service provider automated basic inquiries, allowing support staff to focus on complex customer needs, leading to closing on cases and clients more quickly and an increase in engagement and time spent with their support reps when they need it the most.
   - A healthcare company reduced patient registration time from 10 to 3 minutes by improving both patient satisfaction and staff morale.
   - A bank automated enabled self-service for basic transactions so relationship managers could focus on high-value advisory services.

4. Accelerate company integration

- A global manufacturer cut acquisition integration time from 12 months to 90 days by standardizing their onboarding process into well-defined systems and processes.
- A retail chain reduced new store setup time from 6 months to 6 weeks through automated systems deployment.
- A technology company created a repeatable integration playbook and standardized data structures across divisions that cut monthly migration and reporting costs by 40%.

5. Optimize resource allocation
- A logistics company automated data entry processes, redirecting staff to customer relationship building and inside sales activity that increased sales volume.
- A healthcare provider initiated self-service scheduling, allowing staff to focus on patient care instead of administrative tasks.

The key principle here is that digital transformation means empowering people to do more valuable work. Organizations must clearly communicate that automation's goal is to enhance human capability, not replace it. When employees understand that the goal of the transformation is to help them do more meaningful work rather than threaten their jobs, they become partners in transformation rather than resistors of change. Many times, employees believe technology is intended to replace their work or that they lack the skills required to engage in higher value tasks. Understanding these employee fears and how they dampen motivation can help you adopt a communication framework and the right goal setting and communication project posture that will lead to successful integration of change across the organization.

Also, when workers see their personal values reflected in the company's mission, they become advocates for change rather than just another employee. Such alignment transforms resistance into enthusiasm, and what could have been an obstacle now becomes an opportunity for growth. The key is to make that unification of organizational purpose and personal fulfillment both tangible and meaningful at every level of the organization.

To assess your organizational purpose and values, ensure clear communication on the following:

- Core mission and values
  - What truly differentiates your organization?
  - Which aspects of your operations are non-negotiable?
  - What unique value do you provide that technology should enhance, not replace?
- Market position
  - How do customers actually interact with your organization?
  - What invisible processes make your organization effective?
  - Which competitive advantages must be preserved?
- Strategic objectives
  - Are your goals truly measurable and valuable?
  - Do all stakeholders share the same definition of success?
  - How will you measure progress beyond the technical metrics?
- Cultural identity
  - What unwritten rules govern your organization?
  - How do decisions really get made?
  - What cultural elements must be preserved through transformation?

Before any technical implementation begins, organizations must understand their true posture and identity—not just their processes, but their purpose. They must know which elements of their operations truly create value and which human factors make their organization unique. Only then can they design implementations that enhance rather than erode their competitive advantages.

Leadership Posture: Mindset and Approach

Success in digital transformation requires much more than just choosing the right software, setting good goals, and cultivating the right people, all of which are needed just before the kick-off. What's crucial as the implementation progresses is resilience—the ability of a business to be flexible as it goes about overcoming various difficulties along the way. This should be expected and accounted for in the process. A culture and leadership posture that supports continuous learning and improvement throughout the project are to be valued.

This leadership posture calls for those who can understand and respond to the psychological dimensions of change and the anxiety that comes to an organization learning new systems. The fear of becoming obsolete and the stress of adapting to new workflows call for leadership that can balance empathy with clear direction—empathy for the emotional responses, yet firm in maintaining momentum toward strategic goals. It's about creating psychological safety while pushing for necessary evolution.

Here is an example showing how understanding and addressing the psychological dimensions of change can mean the difference between transformation success and failure.

Fictional Example

Something was wrong. The Operations Director felt it before he even looked at the numbers.

The phones barely rang anymore. The steady rhythm of conversations, once the heartbeat of the customer service floor, had slowed to a trickle.

His gut tightened as he studied the latest sales report—a 30% drop in three months.

"I don't understand," he muttered, turning to the project team gathered in his office. "The portal launch was perfect. Every technical benchmark hit. Zero downtime. Yet our sales have dropped 30% in three months."

On the conference screen, the new customer portal shone brightly—a work of art, including modern web design, sleek interface, strong search capabilities, and real-time inventory, all seamlessly integrated with the backend system. The multimillion-dollar investment had delivered exactly what had been promised: a state-of-the-art e-commerce platform for classic car parts.

"Look at these features," the IT Director emphasized, scrolling through the portal. "Advanced filtering, 360-degree part views, detailed engineering specifications. Everything a customer could need—"

A commotion outside interrupted him. Through the office windows, they saw the company's top sales representative engaged in an animated phone conversation.

"No, I completely understand," he was saying, scribbling furiously on a notepad. "The portal shows the '67 Mustang fuel pump as compatible, but with your modified carburetor..." He paused, listening intently. "You're right—that would cause vapor lock issues. Let me recommend an alternative setup..."

The project team sat in silence and watched him take the next twenty minutes to walk his customer through options, various installation approaches, and related insights from similar restoration projects.

"See what I mean?" The Customer Service Manager gestured toward him. "This is the third client today who tried to use the portal and ended up calling us anyway. They don't just want parts—they want a community of car geeks."

The Operations Director picked up a stack of customer feedback forms. "Listen to these comments: 'Miss the helpful advice from the sales team.' 'Can't discuss compatibility concerns online.' 'No way to ask about alternative approaches.' We didn't just sell parts—we were selling a community based on expertise, relationships, confidence."

He walked to the whiteboard and wrote a striking statistic: "82% of our orders before the portal involved consultation with our sales team."

The room fell silent. Their attempt at digital transformation had inadvertently stripped away their core value proposition. The portal wasn't just a new sales channel—it had fundamentally changed how customers interacted with the company's most valuable asset: its experts.

"We automated the wrong thing," the Operations Director concluded. "We thought we were in the parts business, but we're really in the expertise business. The parts are just the medium for delivering that expertise."

The Manager of Customer Service leaned forward and nodded. "Our customers aren't simply buying engines, fuel pumps, and carburetors. They're buying decades of restoration experience, relationships with experts who understand their projects, and the confidence that comes with professional guidance."

A week later, the project team was back in front of the business with a different approach: instead of replacing human expertise with digital efficiency, they would use technology to enhance it. The portal would be redesigned to facilitate inside ordering, not focus on self-service. The crucial connection with customers and the company's experts would be enhanced not removed.

It would include video chat integration, expert profile pages, the ability to schedule consultations for projects, and collaborative project spaces. This technology would extend the reach and speed of their experts, not replace their expertise and the relationship with the customers.

Success requires understanding the true value proposition beyond the surface-level transactions and recognizing when technology should augment rather than automate. It is essential to preserve the critical human elements that drive business success while designing solutions that strengthen, rather than sever, customer relationships. There is, of course, a place for full automation and self-service capabilities, but the most effective systems amplify unique expertise rather than trying to replace it. This approach ensures that technology serves as a tool to support and enhance the value that human professionals already provide.

Successful organizational posture requires cultivating four essential mindsets across the organization:

- Resilience in facing challenges
    - Acceptance that setbacks are normal and expected
    - Commitment to finding solutions rather than assigning blame
    - Ability to maintain momentum through difficulties
- Adaptability to change
    - Openness to new ways of working
    - Willingness to question established practices
    - Flexibility in approach while maintaining core objectives
- Goal oriented focus
    - Constant evaluation of changes through customer impact lens
    - Balance between standardization and customer needs
    - Commitment to enhancing rather than degrading service levels
- Continuous learning commitment
    - Recognition that implementation is an ongoing journey
    - Investment in regular skill development
    - Culture of knowledge sharing and improvement

While mindset and approach set the stage for transformation, organizations must also establish repeatable practices that reinforce and sustain change. Let's think about how creating a culture of repeatability ensures lasting success.

Budget and Timeline Posture: Rhythm of Repeatability

A digital transformation project that has an official end date on the calendar is destined from the beginning for failure—a magical date where everything will be done and the organization can breathe a collective sigh of relief because they can stop having system implementation meetings, design sessions, and status reports. Successful transformation requires establishing a cultural heartbeat, a rhythm of practices and behaviors that sustain change long after the initial launch. This pattern of activities will consciously reinforce a culture of values and goals within an organization and offer a sense of shared identity. As you know from the sections above, these are vitally important and not just another item to check off a to-do list.

A great transformation never ends but just moves to the next core focus, key goal, and objective of the organization. Organizations should only be finished with transformation when the business they are operating ceases to innovate. However, the rhythm of repeatability comes when the organization innovates, teams align, and technology brings solutions in a consistent flow of successful achievement of goals. Ask yourself the question, "Why would I want innovation and added efficiency to be considered finished? Why would an organization want to stop adding value for clients, employees, and departments?" Is it because you don't have the right staff in place with the correct priorities to keep up the intensity of innovation? Is it because you do not have the budget to keep technology staff, consultants, and systems changing? Is it because the technology goals are not aligned to business objectives that are measurable and drive revenue or expenses in the right direction? Or, is it because the value expected from these initiatives was never actually achieved? These reasons should be a challenge to solve the underlying organizational issues not to simply stop transforming or moving forward with new technology initiatives.

When organizations lack a culture of repeatability, they run the risk of quickly accumulating technological debt, and future transformation becomes more difficult, expensive, and more painful for the organization. It will become urgent, and when it does, it becomes more difficult to justify the organizational costs associated.

A culture of repeatability can take many forms. These range from regular team meetings aimed at enhancing performance to company-wide celebrations of technology changes and of anniversaries, which build team cohesion and motivate employees toward a new goal. Establishing quarterly goals and annual measurements for change and technology initiatives can create a constant posture of repeatability.

The above-described patterns are touchpoints that provide operational structure and create psychological anchors that help people navigate change. When done thoughtfully, these patterns will transform abstract organizational goals into tangible, shared experiences that energize and align teams.

Here are the key principles to keep in mind when building a culture of repeatability:

- Start with small, high-frequency patterns
    - Begin with daily activities that take 15 minutes or less.
    - Focus on immediate value delivery items.
    - Build momentum through quick, goal-oriented wins.
    - Create visible progress markers toward the goal.
- Embed in existing workflows
    - Integrate technology with current department routines.
    - Embed technology service providers within existing departments.
    - Leverage existing communication channels.
    - Minimize or compensate for additional time commitments.
- Ensure clear value connection
    - Link added activities to personal value improvements.
    - Demonstrate real time-saving benefits.
    - Share concrete success examples with the organization.
    - Measure and communicate impact to teams.

- Build mutual accountability
  - Create peer support networks.
  - Establish clear role responsibilities.
  - Track and share progress transparently.
  - Celebrate team achievements publicly.

Consistent repeatability within the technology of an organization is about capabilities and habits. With a strong foundation of repeatable technology practices in place, organizations must next assess and develop the capabilities required to execute and sustain transformation. Understanding your true internal capabilities, both present and future, is essential for successful repeatability. Having built a team of people who thrive under change and constant innovation is key.

Here are some ways that you can assess capabilities:

- Start with honest evaluation
  - Assess team skills objectively.
  - Identify real vs. assumed capabilities within teams.
  - Measure against implementation needs.
  - Document the capability gaps clearly.
- Build systematic improvement plans
  - Prioritize critical capabilities.
  - Create realistic skills development timelines.
  - Allocate adequate resources.
  - Track progress consistently.
- Address root causes
  - Look beyond surface symptoms of capability gaps.
  - Understand capability dependencies.
  - Fix fundamental weaknesses first.
  - Build repeatable education solutions.
- Maintain focus on sustainability
  - Develop internal self-sourced training capacity.
  - Create knowledge retention systems.
  - Build redundancy in order of critical systems.

Building capabilities provides the foundation for transformation. Organizations must also understand and shape the habits that drive daily operations. Organizational disciplines are the repetitive practices and activities that have been ingrained into the team, done almost reflexively.

Organizational habits are deeply human constructs, shaped by countless individual decisions and interactions over time. These patterns of behavior reflect not just formal processes, but the underlying values, relationships, and unspoken norms that guide how work really gets done. Understanding and intentionally shaping your habits requires recognizing that people aren't just creatures of logic alone; they're beings of emotion and routine who need compelling reasons and consistent support to change established patterns. Success in repeatability demands nurturing habits that align both the rational and emotional aspects of human behavior with organizational goals.

- Communication habits
  - Regular status updates and reporting structures
  - Documented communication protocols
  - Official problem-solving channels
- Problem-solving habits
  - Early issue identification
    - Early warning systems and escalation mechanisms
    - Problem recognition patterns and responsible individuals
    - Documented escalation methods and communication chains
  - Solution development
    - Brainstorming practices and mind-mapping exercises
    - Innovation patterns and decision-maker definitions
    - Resource allocation and prioritization matrix
    - Managing the chain of custody for requirements
- Implementation habits
  - Execution and continuous release cycle protocols

o   Follow-through and continuous testing

o   Success measurement practices

o   Learning and after-action capture methods

Making the Most of the Humans in the Loop

Most digital transformation project cycles are between 90 days and 24 months. That is a lot of time and can amount to sometimes 25% or more of careers within an organization. If you assume you will have the same staff at the kickoff of a project that you do at launch,that no one will have left and personal issues won't impact capacity, you are clinging to fool's gold.

A typical employee will easily spend more than half of their waking day and life at a job or pursuing a career. For those who travel consistently for work, this number can be much higher. Our career is not something apart from our life; the concept of complete work-life separation is an illusion. A great or not-so-great day at work translates into how one interacts with friends and family, and stress at home affects productivity and success at work. The goal shouldn't be to remove these effects, or strive for the mythical work-life balance, but rather to acknowledge and deal with them openly.

Take this as an example: a colleague was underperforming, and despite multiple conversations, they never shared openly what was wrong. Their responses were always vague—"I will do better" or "I will work harder"—yet their performance remained uncharacteristically poor. Eventually, they were fired, and only then did they openly show emotion, revealing personal struggles with relationships at home, health issues, and family pain. The real issue was never about their work production or passion but about challenges they had been hiding. The truth is, when something is wrong, those closest to us can often sense it. Instead of suppressing struggles, openly acknowledging them allows for better support and solutions.

The pressure to appear unfazed comes from something deep in our professional DNA. We build our careers on being seen as capable and in control. Think about your last project meeting. How many people admitted they were struggling? Probably very few. We've all been conditioned to believe that showing any crack in our professional armor could damage our career prospects or make us seem less competent.

This mindset gets reinforced by workplace dynamics. In most organizations, there's an unspoken competition for promotions, recognition, money, and influence. Showing vulnerability feels risky when we know others might use it against us. We've all seen or heard stories of someone being passed over for an opportunity because they once admitted they were overwhelmed.

That said, vulnerability in the workplace must be balanced. It shouldn't come across as overwhelming negativity but rather as clear communication: "I may not be at 100% right now due to personal challenges, but I will give my best with the capacity I have, and I'll be back to full strength soon." This approach fosters understanding, maintains professionalism, and ensures both personal well-being and team success.

At its core, it's about basic self-preservation. Nobody wants to be seen as replaceable or less capable than their colleagues. It's easier in the short term to push through and maintain the image of the tireless professional than to risk admitting we need help. We protect this self-image fiercely, often at great personal cost, and it can have a damaging effect on the projects we are involved in.

When personal issues arise and are not accounted for within projects, the consequences can be far-reaching. Team members may seem disengaged, unproductive, or even combative, leading to misconceptions about their competence or commitment. What seems like a lack of communication or poor posture in a professional context often masks deeper, unresolved personal struggles. The result is not only suboptimal project outcomes but also strained relationships and diminished morale among team members.

Addressing these hidden project issues requires a culture of open communication and supportive posture. When employees feel safe to express their vulnerabilities without fear of retribution, it allows for timely interventions and appropriate support. By integrating structured check-ins, multiple channels for raising concerns, and clear guidelines for appropriate disclosure, organizations can ensure that personal challenges are acknowledged and managed, fostering a more resilient and cohesive team.

Rather than attempting to separate personal and professional lives, organizations that succeed at digital transformations are adopting the following models for people on their teams:

- Clear communication frameworks
    - Structured check-ins that go personal and beyond project status
    - Multiple channels for raising personal challenges beyond just HR
    - Support pathways that respect privacy while enabling assistance
- Capacity management protocols
    - Flexible workload adjustment mechanisms
    - Temporary role redistribution processes
    - Clear communication templates for capacity changes
- Leadership response guidelines
    - Training for leaders on how to manage personal disclosures
    - Balance between empathy and accountability
    - Recognition that the best-performing staff produces the best project outcomes

The human element in projects depends on understanding and addressing the complex psychological dynamics at play within teams. From the anxiety of learning new systems to the interpersonal tensions that arise during periods of change, emotional factors profoundly impact project outcomes. Leaders must develop high emotional intelligence, rigorous habits, and a firm grip on outcomes and objectives to recognize when technical challenges mask deeper human concerns and when resistance signals unaddressed fears.

As we move forward in subsequent chapters, remember that digital transformation is fundamentally a human journey enabled by technology, not the other way around.

In reality, success is based on understanding and addressing human needs prior to technical solutions, building organizational capabilities supportive of sustaining change, and fostering cultural environments that embrace transformation.

Change is not about the systems, but about people. The most sophisticated technology will not overcome human resistance, and quite often, the most elegant solution will fail to deliver results when there's no buy-in. Understanding and embracing that people are at the center of digital transformation is the key to creating sustainable change that will deliver value that lasts.

Reflection Questions

1. How does your organization currently handle the intersection of personal and professional challenges? Does turnover of resources affect your projects?

2. What are some of the "technological debts" that your organization has put off because they are perceived as too expensive, challenging, or time-consuming?

3. What informal workflows or habits might be invisible to your formal processes?

4. How do your current change management practices address human psychological needs?

5. Does your organization view the completion of digital transformation as a date on the calendar? Are there budgets in place to continually innovate?

6. How does your leadership team balance empathy with accountability?

CHAPTER 3

Rebuilding the Foundation: Before and After Project Failure

> *"The best time to plant a tree was 20 years ago.*
> *The second best time is now."* —Chinese Proverb

If you are reading this, it is likely that you are about to embark on a digital transformation project like an ERP implementation, are currently mired in one that is failing, or have previously been part of one that did not go well and want to make sure not to repeat the mistakes of the past. In these types of rescue situations, most companies get stuck in a vicious cycle of blame, finger-pointing at vendor choice, technical problems, or resource limitations. Perhaps that is the root of the unfortunate and sobering statistic that 60-80% of technology deployment initiatives fail to achieve their goals (Bojinov, 2023).

When failure occurs, deadlines shift, employees become overworked, and issues escalate beyond what the business anticipated. A direct consequence of these issues is unanticipated cost increases which create additional pressure and stress on all parties involved. Acknowledging the symptoms of a troubled project is the initial step toward recovery. If these serious symptoms occur, it's time to stop and rebuild, not from where you are, but from first principles. The objective in such cases isn't to save what you've created, but to make what you're attempting simpler. As Elon Musk famously recommends, one should take everything back to the bare essentials and then reassemble. In fact, Musk asserts that "if you do not end up adding back at least 10%, then you didn't delete enough" (Minnaar, 2023). This maxim is exactly applicable to technology implementations that have derailed.

Simplification needs to occur in three main dimensions: team structure, technology strategy, and, most importantly, processes. Automation of processes has been considered the bread and butter of digital transformation. In this way, businesses rush to automate their organizations based on misguided advice from internal stakeholders or third-party vendors who give assurances of efficiency, accuracy, and cost savings. Yet for every successful automation project, two others fail at considerable cost. What companies don't fully understand is that when you go out and purchase new software, you're not purchasing a solution first; you're purchasing a mirror that will reflect back the reality of your processes, your people, and your problems. This reflection, sometimes unflattering, is where so many automation projects fail, not because of technology, but because a solid foundation wasn't built.

The reality of digital transformation is that you can't automate chaos and receive order. When companies hasten to automate, while lacking solid, documented processes and people with deep understanding, what they most commonly receive are costly systems that will illuminate their issues instead of resolving them.

A Company Example

The open office area of the AP department, which ought to have been abuzz with staccato keyboard clicks and shuffling invoices, had turned unsettlingly quiet. The sole sound that pierced the silence was the whine of the automated scanner—a $1.2 million expenditure in technological excellence, now a symbol of organizational arrogance.

The AP Manager's mouse lingered over a dashboard of real-time invoice processing metrics, each red indicator highlighting an error that occurred. Six months after launching their "revolutionary" AP automation platform, the situation had reached a boiling point. The inbox was overflowing with error notifications: angry suppliers threatening to stop deliveries, executive inquiries regarding payment status, and a meeting invitation from the CFO to the department with an ominous subject line: "AP Emergency Response."

"This can't be right," muttered the AP Manager, contrasting before-and-after metrics. What had been promised as a 75% decrease in manual labor had somehow been converted into a 50% increase in overtime labor. Average payment processing times had doubled.

Down the hall, a 20-year AP veteran pounded her fists on her desk. "The system's doing it again," she yelled, exasperation in her voice. "It's kicking back every construction invoice transaction as a duplicate because it can't differentiate between the PO numbers and invoice numbers. We're going to lose our supplier if this continues. Someone needs to manually process these invoices and that will take days to complete."

Across the room, another expert multi-tasked with three screens—one displaying the new automated system, another with their old manual tracking spreadsheet, and a third with vendor emails requesting payment status. "Remember when we could just look at an invoice and know what to do with it?" The rhetorical question hung in the air, weighted with irony.

...

Later, in the glass-walled conference room, the emergency CFO meeting unveiled a reality growing more complicated by the minute. As laptops snapped open and screens flickered to life with data, the system implementation consultant hovered near the whiteboard, marker at the ready. "Take me through exactly how an invoice gets processed. Not the optimal flow—the real one."

What emerged on the whiteboard over the course of the next hour was not the neat, linear progression that had been envisioned.

Rather, a spiderweb of workflows, with each line denoting a route that an invoice might follow.

Construction invoices had to be specially coded for work sites. Grants required additional approvals. Some vendors used EDI, some PDFs, but others wanted scanned paper copies with wet signatures. Special payment terms had been negotiated by certain departments. Fiscal year-spanning projects required special treatment. The so-called simple AP process, it turned out, had dozens of one-off situations, exceptions, and rules of thumb that were causing the errors. This complexity was only revealed when a system attempted to automate it.

"Look here," the new consultant indicated a very tight clump of arrows. "Your staff processes fourteen various invoice formats from your major vendor alone, with different rules for each one. The software is not the issue; no one ever designed for these various invoices and rules. It is revealing how fragmented the process was all along."

The CFO leaned forward, coffee untouched. "But it worked before. We got payments out."

"It worked," the consultant replied, "because your team was silently managing this complexity. They were the human middleware, translating between different systems, different expectations, different needs. We didn't automate your process—we automated your chaos."

The AP Manager leaned back, the gravity of the situation taking hold. They'd invested seven figures in attempting to optimize a process they'd never truly understood. The software wasn't the answer, it was the messenger, bringing unwanted realities regarding their organizational blind spots that clever tenured employees had worked around for years.

"The way forward," the new consultant went on, putting the lid back on his marker, "isn't through more automation, at least not to begin with. We must make the invisible processes and rules visible with some labels. Record every variation in the process. Chart every vendor exception. Identify the cause of every special case. Only then can we design automation that assists instead of obstructs."

Three months later, the AP department hummed with a new vibrancy. Gone was the panicked juggling of multiple systems and spreadsheets managing the exceptions. The automation platform, rewritten to mirror their real-world workflows rather than some idealized interpretation, had at last begun to live up to its promise. But the real transformation wasn't in the software; it was in their understanding of their own internal processes.

"The ultimate irony," the AP Manager later remarked in a conference presentation, "is that we thought we were purchasing a solution, yet what we actually purchased was a mirror. Sometimes you need to see the mess in clear focus before you can effectively repair it."

In the back of the room, there were nods of recognition. It was a story being played out in offices everywhere: automation does not fix broken processes—it amplifies them. The key to a successful journey toward automation is not in implementation of technology itself, but in the painstaking effort of understanding, documenting, and optimizing the human systems that automation is intended to supplement or replace.

The lesson was straightforward. Before automating, illuminate. Before changing, comprehend. And most of all, acknowledge that the greatest asset in any business is not computer programs, it's the human understanding behind the software that solves problems and builds relationships.

Through both successes and failures in guiding technology implementations, a crucial lesson has emerged: rebuilding the foundation after project failure requires more than just technical fixes. It demands a systematic approach to understanding and addressing the core issues that caused the failure in the first place.

In this chapter, we'll examine proven strategies for both preventing implementation failures and rescuing troubled projects. We'll explore how to:

- Break down complex initiatives into manageable components
- Establish clear roles and responsibilities that stick
- Build genuine team alignment and sustained momentum
- Create effective decision-making frameworks that prevent project derailment

We will look at real-world examples of organizations that have successfully navigated these challenges, turning potential disasters into successes. Because while preventing failure is ideal, knowing how to recover from it is equally crucial in today's rapidly evolving business landscape.

Critical Questions That Often Reveal Readiness Gaps

When organizations express interest in automation, the first response from many software vendors is to schedule demos and discuss technical features of their product. This is exactly backwards. The critical first step should be assessing organization and process readiness through a series of foundational questions that often reveal the dangerous gaps in organizational maturity.

Consider our AP automation example. In working with hundreds of organizations, we've found that three key areas consistently determine success or failure. They are discussed below, with the addition of some of the probing questions that should have been asked in this specific AP example:

1. Data integrity

   "Irma in accounting handles all our vendor setups," the AP supervisor explained proudly during a readiness assessment. "She's been here twenty years and knows all the special payment terms and requirements."

This response, while common, reveals a critical flaw. The entire vendor management process relies on individual knowledge rather than documented procedures and system-enforced controls. When pressed with specific questions, the gaps become clear:

- Are all vendor records complete and up to date?
  - "Well, Irma updates them when she can based on the emails sent from the project managers..."
- Do they contain accurate ACH details?
  - "Some do, but others are in a spreadsheet and emails on the shared drive."
- Are these details stored in a standardized format?
  - "All departments have their own way of sending them."

Recommended Next Step: Perform an exhaustive data audit to determine discrepancies, missing data, and quality problems prior to the selection of any automation tool. Define specific data standards and engage in a project for cleansing current records in the existing systems first.

Common Mistake to Avoid: If you hear something like this: "Our software has data migration tools that will automatically cleanse your data during implementation"—beware. Vendors who assure you their import wizards or AI tools will automatically repair years of poor data practices are making false claims. Without defined rules, no software can tell you which one of your three different vendor records is the true record or to which project that vendor belongs.

2. Approval workflow clarity

Similar issues emerge when examining approval processes:

"Each department kind of does their own thing," the controller admitted. "Marketing needs the VP to sign off on everything, but Operations has different rules, depending on the amount and the project. And sometimes project managers just walk around getting signatures on the PO when they need something rushed."

The lack of standardized approval workflows means that automation will either:

- Force everyone into a rigid process that doesn't match business needs, or
- Require so many exceptions and special rules that the system becomes unmanageable

Recommended Next Step: Map your current processes exactly as they are currently operating, not how they are supposed to run in theory. Identify variations and exceptions, then deliberately decide which to standardize and which to leave alone based on complexity, frequency, and importance.

Common Mistake to Avoid: A suggestion like this one right out of the gate: "Let's use the software's industry best practice workflows rather than document your existing ones" is a red flag. Although enticing, forcing your organization into off-the-shelf best practices without an appreciation of your specific requirements often leads to resistance and workarounds. Even if this is the agreed upon end result, doing so without understanding the impact of the change across the organization and getting buy-in from the various process owners will lead to disaster.

3. PO and payment tracking

Few things are more impactful to an organization as the state of purchase order, cash flow, and payment tracking:

"Oh, we definitely use POs," the procurement manager assured us. But further questioning revealed:

- Only certain vendors require POs.
- PO numbers are often created after the fact to match invoices.

- All project administrators had the ability to issue new POs to a job without signed agreements.
- No consistent process exists for matching POs to vendor invoices by job.
- Payment status updates happened manually during a monthly reconciliation by the accounting team and included manual adjustments to the job so budgets matched the PO and invoice without understanding why there is a difference.

Recommended Next Step: Normalize vendor data input with precise standards and automatic checks. Implement PO and job discrepancy approval processes in the current system and prohibit the editing of transactions without reason codes. Audit data records on a regular basis to achieve accuracy and completeness.

Common Mistake to Avoid: "The software will help us implement rigor and standard procedures after the implementation" is a misleading claim. You cannot rely on software alone to fix issues or count on standardization being addressed after implementing the technology. Automation is helpful, but many times human oversight is required to maintain system integrity.

These glaring gaps make automation impossible, and they also make it dangerous. Systems are based on code, and code runs on rules, equations, and deterministic patterns. When you try to automate processes that are full of exceptions and variations, you're trying to get the system to do something it is inherently unable to do. A computer program cannot understand "except when Melissa calls," or "only at quarter-end," or "just for our big customers" unless programmed specifically for each circumstance. Without clean data, consistent workflows, and normalized processes, automation doesn't merely trip; it crashes and burns, amplifying chaos instead of efficiency. The system will relentlessly perform precisely what you've instructed it to do, whether or not that's what you meant to happen.

Why Process-Blind Automation Fails

When organizations rush to implement automation systems without addressing these foundational issues, three major problems inevitably emerge:

1. The implementation becomes a costly maze

    "We estimated six months for implementation," an IT director shared during a project rescue assessment. "We're now eighteen months in, have spent triple the budget, and still can't go live."

    This common scenario occurs because organizations fail to recognize the vast chasm between the goal of a "data visibility system" and an "automation system." If data visibility systems do not exist, then:

    - Significant time and money must be spent cleaning and structuring data that should have been standardized beforehand.
    - Complex workflows need to be documented and coded, often revealing data conflicts that require executive intervention.
    - Integration points with other systems prove more complicated than expected due to inconsistent data formats.
    - Testing cycles repeatedly fail as edge cases and exceptions surface one after another.

2. The system impacts critical business operations

    A manufacturing firm learned this the hard way when their recently automated AP system started making payments.

    They based their automation on the industry-standard "three-way match" process. This is a common practice that compares "the purchase order, invoice, and goods receipt to make sure they match, prior to approving the invoice" (Blaney, 2024). In theory, this should have generated bulletproof accuracy. In practice, it generated chaos:

- Invoices were denied for decimal discrepancies (a $100.00 PO and a $100 invoice) that had previously been determined to be a match by humans.
- Vendors received incorrect amounts because their negotiated payment terms and discounts weren't properly documented and handled in the system, leading to required refunds and back-dating transactions to make things correct.
- Critical production material remained unprocessed due to receiving documentation containing different part numbers than purchase orders (i.e., missing dashes, leading zeros, sub-assembly parts), making the three-way match appear invalid. Emergency payments couldn't be made since the three-way match workflow of the system couldn't support emergency processes that previously skipped steps.

"We had fewer mistakes with the manual process," their AP Manager conceded. "Our veteran staff members knew when to use judgment, such as what price differences were acceptable or which vendors required special treatment. The automated system could only do what it was programmed to do, not use judgment."

3. Shadow systems proliferate

Perhaps the most telling issue is when teams begin creating manual workarounds to bypass the automated system:

- Spreadsheets tracking the actual payment status
- Manual approval emails running parallel to the system approvals
- Paper files for special cases that don't fit the automated workflow
- Email chains documenting exceptions and overrides

"We're actually doing more work now," one AP clerk explained. "We have to maintain the automated system and keep our manual tracking because we can't trust the system to catch everything."

One of the most critical and commonly ignored forces of digital transformation is the likely addition of operational complexity in the implementation process. New systems, particularly complicated enterprise-level systems such as ERP software, will more likely demand additional human interaction, at least in their early launch stages. Companies can find themselves supporting both the new system and legacy manual tracking systems for a period of time, literally doubling up efforts. This is due to the fact that:

- New systems on day one may lack the various exception handling needed.
- Staff must verify outputs from the system against their experience and current processes.
- Complex configurations require elaborate human control and validation.
- Transition phases need concurrent monitoring to generate reliable information.
- Humans using new systems need time to develop muscle memory and will naturally be slower to perform standard processes in a new system until they get the repetitions needed for speed.

Counterintuitively, most businesses first may require more employees, not fewer, to run new systems at optimal efficiency. This seemingly illogical reality requires digital transformation initiatives to expressly recognize heightened operational complexity, heightened training time, and possible near-term productivity loss.

For digital transformations to succeed, organizations must allow this process the time it requires to be done correctly, resisting pressure from arbitrary timelines dictating when staff can be reduced or return to their former routines; only through deliberate, thoughtful pacing can the foundation for lasting change be secured.

Successful rollouts accept that the journey to efficiency is seldom linear, and businesses need to design flexibility and added support into their transformation initiatives.

Building the Right Foundation

Success in automation requires inverting the traditional approach. Instead of starting with software selection, organizations must first:

1. Standardize data management
   - Create and enforce data standards
   - Clean and verify existing data
   - Establish processes for ongoing data maintenance
   - Define clear ownership of data segments
2. Document and optimize workflows
   - Map current approval processes across all departments
   - Identify, document, and resolve process variations
   - Create clear documentation of standard procedures
   - Define and test exception handling
3. Establish control mechanisms
   - Implement tracking for all related documents
   - Create reconciliation procedures
   - Define clear audit trails
   - Set up monitoring and reporting systems
4. Validate through testing
   - Run pilot processes with real transactions
   - Test exception scenarios
   - Verify reporting accuracy
   - Confirm user acceptance

Only after these foundations are solid should organizations proceed to the next step of automation. This approach might seem slower initially, but it dramatically increases the likelihood of success.

When an implementation has gone off the rails due to premature or misguided automation, the most counterintuitive yet often most effective strategy is to step back. This means deliberately dismantling automated processes and returning to manual, foundational workflows. It's a strategic retreat that allows organizations to rebuild their operational core before reintroducing systematized automation solutions.

Breaking Down Complex Projects

Despite good fundamentals, big change efforts can also falter without an effective mechanism for managing complexity. Most organizations fail to document all of their system requirements in a way that can provide the details needed to prioritize, define the outcomes, and then actually implement the solution. Here's an example from a worldwide manufacturer:

"We're six months into our ERP implementation," the project manager described, "and we have 8,000 documented requirements in our tracking system. But no one can tell me which ones are really critical to go-live versus nice-to-have. No one knows how the requirements are interdependent. We're getting lost in the weeds and losing sight of the big picture."

This is where most companies get stuck. Organizations all too frequently dive into detailed requirements without creating a clear hierarchy of needs. What happens? Analysis paralysis, scope creep, and distraction from essential objectives. Many organizations will do the initial work of documenting the highest-level requirement. After that, some do a good job on the detailed technical specifications, but most miss all of the in-between. They miss:

- Why is this requirement valuable?
- Who is the owner of the workflow?
- What are the anticipated outcomes?
- When is the right time to implement that requirement?

Organizations need to engage in the middle for their requirements, provide the context to the organization, and have a system of documentation to which the entire digital transformation team adheres. Somebody must be named responsible for holding the entire organization accountable so that the process of documenting requirements is not circumvented.

Although numerous methodologies exist (i.e., Agile, Waterfall, SAFe), we've discovered that complicated initiatives require a simpler, more natural methodology. For instance, the use of a simple requirements traceability matrix (RTM) is a deliberate documentation technique that changes a normal laundry list of requirements into an indispensable project management instrument (Sire, 2024). In substance, the RTM delivers an exhaustive framework that records every requirement's distinct identifier, clear description, source, priority rank, status, and final acceptance criteria. In contrast to conventional requirement tracking, this method delivers a functional document that is more than a list of technical specifications. It identifies the underlying business value, wished-for outcomes, and key interdependencies.

This traceability is the cornerstone of project governance, scope control, and test approach, and ultimately makes successful realization possible. By providing clear connection from strategy to delivery, organizations make sure that technical delivery is in support of business requirements instead of being an end in itself. Though the RTM does make a powerful tracking tool, it remains at best potential only with the complement of a structured, documented, and detailed requirements-gathering framework. For this example, we will use the Level 1-6 Requirements Framework. This framework, developed by Dynamic Consultants Group (DCG), provides a comprehensive approach to systematically defining and categorizing project requirements with the primary objective of simplification.

A Simplified Requirements Framework

Level 1: Strategic Objectives
- What business outcomes must this project deliver?
- Which problems must it solve?
- How will success be measured?

Level 2: Core Process
- Which business processes are in scope for that objective?
- How do these processes interact?
- What are the critical dependencies?
- What technology will be used to facilitate the process, if any?

Level 3: Functional Requirements
- What specific capabilities are needed to facilitate each process?
- Which user groups are involved?
- What are the key integration points?

Level 4: Process Workflow and Scope Assessment
- What is the detailed step-by-step workflow for the functional requirements?
- How will users interact with the system?
- What is the estimated effort and cost? Is it worth the investment?
- Is this sufficient to proceed to detailed design?

Level 5: User Experience Design
- How will each screen and interaction function?
- What are the specific data fields and validations?
- What are the user pathways through the system?
- How will exceptions be handled?

Level 6: Technical Implementation Specifications
- What are the specific development frameworks used?
- What are the systems and data table relationships?

- How will integration points be constructed?
- What are the security, performance, and compliance requirements?
- What architecture patterns will be followed?

Most forward-thinking organizations bring in seasoned partners to lead the strategic framework and governance process, even if a different vendor is selected to perform the technical implementation work. That way, they can have both strategic alignment and technical expertise that hold each other accountable. This is a key ingredient for project success. The most effective transitions are built on models that have been tuned by field experience in many enterprise deployments. In considering implementation partners or vendors, whether or not they can show this degree of rigor and discipline should be a top factor in your selection criteria.

This simplified hierarchical approach ensures that every detailed requirement (Level 6) traces back to a strategic objective (Level 1). More importantly, it provides a clear framework for decision-making. It provides a clearly defined scope and requirements control mechanism. Organizations reap huge rewards using such ordered methodology, one that includes meticulous documentation at every level and guides teams through the tricky implementation process.

Example: A Common Mistake in Requirements

For many organizations where visibility or reporting is a problem for businesses, there's a temptation to point fingers at the system instead of looking at the underlying processes and data practices. This is a trap and creates an expensive, vicious cycle:

1. A business recognizes an issue (typically data visibility, reporting, or automation).
2. Leadership assumes existing systems are not sufficient or uses the issue as a pathway to change in systems or technology.

3. A single large-scale digital transformation implementation is launched.
4. The new system(s) will have the same data and process issues as the last because no one addressed the underlying problem.
5. The implementation does not provide the anticipated value or solve the issues. Many times, it creates new problems or highlights current challenges that were not glaring before.
6. The cycle starts all over again (most of the time with new leadership or a new selected implementation partner).

What's particularly nefarious about this cycle is that it drains millions of dollars in resources without addressing the underlying reasons. It's like replacing your vehicle's engine because the gauges in the dashboard are inaccurate, rather than simply fixing the performance measures. The solution is to step back and implement a framework for gathering detailed requirements.

With requirements properly structured, the next critical element is establishing clear decision-making ownership. This means creating a decision matrix that specifies:

- Who must be responsible for executing the decision?
- Who must be accountable for the outcome?
- Who needs to be consulted before deciding?
- Who needs to be informed after the decision?

For example:

- Strategic decisions (Level 1-2) typically require executive approval.
- Process decisions (Level 3) need business unit owner sign-off.
- Prioritization and workflow (Level 4-5) can often be made by the implementation team as long as additional budget or time is not required to be approved at the executive or business unit owner level.
- Testing and technical design decisions made (Level 6) should involve the IT and technical team.

During the implementation cycle, this level of clarity prevents the common scenario where decisions are:

- Delayed while seeking unnecessary approvals
- Made without consulting key stakeholders
- Reversed because the wrong people were involved

Once decision ownership is settled, we must think about the reality of resource allocation to each level of requirements-gathering.

Building and Maintaining Team Alignment

"We have the budget approved and the decision framework in place," a program sponsor recently told us. "But everyone assigned to the project still has their day job. How do we make this work?"

This touches on perhaps the most overlooked aspect of complex projects: resource allocation.

The most thoughtfully planned requirements and resource allocations are worthless without actual team availability. Take a look at the following example from a recent ERP implementation:

"On paper, everybody was dedicated to the project," the program manager said. "We had all the department managers signed off. But three months in, attendance at critical meetings was slipping. Critical decisions were being postponed. Individuals who were meant to be dedicated to the project were getting pulled back into their regular jobs because of staff turnover, vacations, holidays, and organizational day-to-day demands."

Below are three key principles you can use to guide this question:

1. Capacity planning must be realistic.
   - No one can give 100% of their day to the project. This includes your consultants and vendors.
   - Plan for a maximum of 50% availability from internal key resources.

- Account for vacation, holidays, sick time, and other commitments.
2. Skills must match responsibilities
   - Don't assign technical decisions to business users.
   - Don't burden developers with process design.
   - Ensure project roles align with actual capabilities.
3. Identify backup resources with decision-making authority.
   - Critical roles need documented backups.
   - Cross-training should start early.
   - Knowledge sharing must be ongoing.

Rebuilding your project's foundation requires building a resilient and effective digital transformation team. This begins with a pragmatic approach to resource assignment that recognizes the constraints of availability, matches skills to responsibilities, and ensures thoughtful backup coverage. Yet, assembling the right people is only the beginning. As the project cycle intensifies and challenges emerge, true success hinges on the team's ability to remain aligned in vision, purpose, and execution.

"Do I have the right people on the team?" This question echoes a common myth that organizational team alignment can be achieved through a one-time team assignment. Real team alignment is the result of understanding and addressing three critical factors of the team from time to time throughout the various project phases. This can be simplified using the GWC framework. GWC is an EOS tool, from The Entrepreneurial Operating System® (EOS®), that enables leaders to identify and solve team member problems quickly and objectively (Breyley, n.d.). Here are the three dimensions of evaluation that should be done for every member of the project team.

1. Get It
   - Does the team member truly understand:
     - The project's purpose and value
     - Their specific role and responsibilities
     - How success will be measured

- o The impact on their daily work
- o The aptitude needed to learn new technologies
- To achieve this, concrete actions are required:
  - o Clear communication of project goals and benefits
  - o Regular reinforcement of key messages
  - o Visible executive support and participation
  - o Concrete examples of how the work will improve

2. Want It
   - Is the team member genuinely motivated to:
     - o Participate in the change
     - o Learn new processes and systems
     - o Take on additional responsibilities
     - o Support others through the transition
   - To achieve this, concrete actions are required:
     - o Personal benefit identification and articulation
     - o Career development opportunities
     - o Recognition and reward systems
     - o Peer success stories and testimonials

3. Capacity for It
   - Does the team member have:
     - o The necessary ability, skills, and knowledge
     - o Adequate time and resources
     - o Required authority and influence
     - o The ability to thrive, not crumble, under pressure
   - To achieve this, concrete actions are required:
     - o Skills assessment and training plans
     - o Realistic resource allocation
     - o Clear authority delegation
     - o Management support agreements

This constant evaluation of the team is a concerted effort and requires diligence to spot issues that arise and fortitude to make swift changes to the team when it is not working. Here are some warning signs to look for in members of the team:

- Consistently asking basic questions about fundamental concepts
- Excessive time spent researching simple tasks
- Lack of understanding of the data or processes they are the owner of
- Need to extensively rework deliverables due to quality issues
- Visible frustration or avoidance when tackling technical challenges

For example, picture a manufacturing firm's rollout of a new ERP system. The process owner possessed good business acumen but little technical knowledge. In planning for integration, they sanctioned an architecture that was not designed to handle the firm's volume of transactions because they did not know the volume of transactions and never asked the question. The mistake was not discovered until UAT. It cost three months and $250,000 in extra expenses for accelerated infrastructure upgrades. Every person on the team must have the get it, want it, and the capacity to be in control of the area of the project they are making decisions for.

"When we began assessing team members in this way," one project director divulged, "we understood why some weren't contributing. Some just didn't understand what we were attempting to do. Others did understand but didn't have a passion for the change, and many simply didn't have the bandwidth to contribute meaningfully to the project."

Cross-department collaboration is another key area to address when considering time commitments from the team. It's also among the toughest to accomplish. Typical obstacles are:

- Territorial protection of processes and data
  - Departments tend to protect their processes, information, and workflows as proprietary treasures. This territoriality is driven by practicality, such as not losing control, as well as psychology, such as desiring recognition for one's status and expertise.

- For instance, in CRM implementation, salespersons resist the sharing of customer relationship information they have developed themselves, and marketing people guard their lead qualification processes, causing tension at key handoff points.
- Different priorities and timelines
  - Each department has different goals and time scales. Production may be concerned with efficiency, sales with revenue, and customer service with response time. These differing interests tend to create tension during system implementations.
- Competing resource demands
  - Finite time, dollars, and attention across the business require that departments compete for resources. When the finance department requires resource availability during month-end close or the warehouse department requires assistance during the peak shipping months, conflicts are inevitable.
- Historical conflicts or mistrust
  - Previously failed projects or interdepartmental conflicts generate ongoing distrust that impedes collaboration. When companies start blaming the last partner or other departments, it is a sign of inherent organizational problems that need to be resolved.

Successful organizations address these through:

- Shared requirements ownership models
  - Joint process design sessions across departments
  - Cross-functional working groups to build the RTM
  - Collaborative decision-making and delegation of authority
  - Mutual success metrics
- Boundary-spanning roles
  - Process owners who cross departments

- Integration specialists who understand workflows across departments
  - Cross-functional coaches or consultants who can bridge the gaps
- Unified communication channels
  - Regular stakeholder alignment forums to ensure the same page
  - Cross-team status sharing and transparency on slipping priorities
  - Joint problem-solving sessions across departments
  - Shared success celebrations that happen publicly

Often, when implementation teams are formed based on organizational titles rather than necessary skills, turf wars erupt. The best transformations establish teams based on capability and perspectives, not titles.

Building Skills-Based Teams

Consider this example from a manufacturing company implementing a supply chain optimization system. The initial project team consisted of directors from each department: Purchasing, Production, Logistics, and IT. Meetings quickly devolved into turf wars as each director defended their department's processes and requirements. Progress stalled for months until the approach was fundamentally restructured.

The transformed team was organized around critical skills rather than departments:

- Process analysis expertise (regardless of department)
- Data management experience, most of the time not in IT but business units
- Change management capabilities
- Technical integration knowledge
- End-user workflow understanding

This skills-based approach resulted in a team that collaborated to solve problems rather than defend departmental interests. The focus shifted from "what my department needs" to "what capabilities our organization requires."

This transition to skills-based teams, however, is usually opposed by titled executives who might feel threatened when decision-making authority moves to those with expertise instead of status. I've witnessed VPs cling to decision-making authority even though their actual expertise was decades obsolete, and I've seen directors who barely understood a subject squirm when experts commanded more respect because they did. Or worse yet, they assumed they knew what they did not, and gave incorrect answers to questions regarding current state processes. It soon became clear they did not know the actual processes that were being used in the very departments they oversaw.

Breaking down such resistance internally takes a fine balance of clarity and respect:

- First, recognize the valid concerns of titled leaders. Their institutional memory, business context, and accountability to the organization continue to be vital. Be clear that a move to a skills-based approach does not reduce their leadership or ability to influence the decision; it augments it by enabling them to coordinate expertise instead of being required to have all the knowledge themselves.
- Second, establish formal venues in which expertise can function without diminishing hierarchy. Investigate the possibility of establishing a decision-making hierarchy in which subject-matter experts have explicitly defined authority over particular decisions while titled leaders maintain authority over strategic direction and resource allocation.
- Third, assist leaders in reframing their value proposition. The most effective executives I have had the pleasure of working with readily acknowledged what they did not know and actively indicated the knowledge that resided in their firms. Their

confidence was not in personal omniscience but in being able to put together and lead the right capabilities at the right moment.

- Finally, codify this expertise-based authority within your project governance. Clarify what must be decided by positional authority (budget, resource allocation) versus by capability-based expertise (technical architecture, process design). Having this clarity reduces the perception that giving power to skilled contributors takes away from leadership positions.

The best transformation creates a culture where everybody understands that modern leadership is not about being an expert in all things; it's about knowing your team and how to manage it.

To compose effective teams that transcend territorial concerns and titles:

- Focus on required capabilities, not representation
  - o Prioritize key competencies over representation
  - o Specify the particular skills and knowledge required to excel
  - o Select team members who possess those skills, regardless of their role in the company
- Include cross-functional expertise
  - o Make sure the team includes individuals who have served across departmental boundaries
  - o Appreciate individuals who understand multiple business operations
- Balance technical and business perspectives
  - o Include both technical experts who understand system capabilities and business experts who understand operational needs
  - o Value team members who can speak both languages
- Consider diverse experience levels
  - o Include both veterans with historical knowledge and newcomers with fresh perspectives
  - o Pair experienced staff with rising talent for knowledge transfer
- Prioritize collaborative mindsets

o   Select team members known for their ability to work across boundaries
o   Value problem-solvers over process defenders

Structured Stakeholder Involvement

Despite clear goals and good team structure, implementation success relies on purposeful stakeholder involvement. Ad hoc stakeholder involvement leads to overlooked requirements, last-minute resistance, and turf battles that nullify progress.

While stakeholders may not desire involvement in every technical detail, it is essential that engagement frameworks provide them with the clarity and context needed to make informed choices at critical junctures. From a consulting standpoint, the absence of structured stakeholder involvement can be perilous; when key individuals opt out of the process, significant risks go unaddressed, decisions are made in a vacuum, and projects may veer off course before anyone notices. Proactive participation, even if only at major milestones, protects the integrity of decision-making and mitigates hazards that stem from disengagement.

If you don't have organized mechanisms of involvement, you only hear from leadership if something is wrong. This creates a vicious cycle where managers gain a reputation for criticism rather than encouragement, appearing to only ask questions or bring disappointment. Stakeholders who speak up only during times of trouble provide input that is reactive, emotional, and out of context for the project. By comparison, regular, thoughtful involvement keeps stakeholders current, enables constructive feedback, and builds the trust necessary to weather unavoidable setbacks.

Organized stakeholder engagement adheres to systematic methods of engaging the right individuals at the right time in the right manner.

Consider, for example, a CEO who deliberately kept their distance during a high-stakes rollout. Their rationale was simple: by staying out of the weeds, they could swoop in at the end, challenge assumptions, and play the bad cop, holding the team to account and pressing for last-minute improvements. Yet, instead of sharpening the outcome, this approach sowed confusion and eroded morale. Key contributors felt their months of disciplined effort were discounted, with priorities suddenly reshuffled and foundational decisions questioned in the eleventh hour. Team members who had poured themselves into the process interpreted the CEO's absence not as strategic but as indifference, and their trust in leadership quietly dissolved. When outcomes didn't match expectations, postmortems revealed the true cost: disengagement from the top meant risks went unflagged, course corrections came too late, and the team's confidence in both the process and their leader was deeply shaken.

Creating Structured Involvement Frameworks

Effective stakeholder engagement requires systematic planning and execution. From the very beginning of a reset or rescue situation, in order to create success, it's critical to effectively deploy the strategy and plan and obtain buy-in from senior leaders:

- Develop a comprehensive stakeholder map
  - How?
    - Identify all stakeholders impacted by or impacting the deployment.
    - Evaluate their levels of interest, influence, and particular concerns.
    - Categorize stakeholders by both impact and authority levels.
  - *Example*: Draw a 2x2 matrix plotting stakeholders versus influence (high/low) and interest (high/low), with particular individuals and groups in each quadrant.
  - Outcome: Steers away from the "where did these people come from?" syndrome when previously unseen stakeholders appear late with essential requirements
- Create tailored engagement strategies
  - How?

- Design different approaches to communication for different stakeholder types and combine communication channels where possible to limit redundancies.
- Establish clear roles: decision-makers, advisors, implementers, and users.
- Match engagement methods to stakeholder needs and project phases.
- *Example*: In an inventory control system, include not just warehouse personnel but also sales (who make commitments against inventory), finance (who estimate its value), suppliers (who stock it), and customers (who receive delivery of it).
  - Outcome: Steers away from the "Why was this decision made without me?" syndrome when previously unseen stakeholders do not feel involved in the decision-making process.
- Implement multiple feedback channels
  - How?
    - Provide structured ways for stakeholders to contribute to the conversations and express concerns in a way that does not cause their team to feel uninvolved in the process.
    - Create both formal (governance committees, review sessions) and informal (office hours, suggestion systems) channels.
    - Ensure channels exist for all organizational levels, from executives to frontline staff, and that they are in attendance at the sessions.
    - If the executives cannot attend, reschedule and ensure attendance. Don't allow for delegation if they are required within a channel of decision authority. Delegation can be used as a way to stay uninvolved.
  - *Example*: Indicate that the CFO has moderate influence but high interest, and is concerned about cost control and ROI,

and that the warehouse manager has moderate influence but high interest, and is concerned with operational effectiveness.

- o Outcome: Facilitates focused communications that speak to the individual stakeholder's concerns as opposed to general updates that don't connect.
- Establish clear decision rights
  - o How?
    - Define explicitly who makes which decisions for which departments.
    - Document and communicate the decision-making framework and hold even the most senior-level leaders accountable to the individuals who hold authority for decision-making.
    - Create escalation paths within the team for addressing conflicts that augment rather than circumvent the overall process.
  - o *Example*: For the busy CEO, hold monthly 15-minute stand-up briefings; for department leaders, hold bi-weekly working meetings; for end users, hold periodic focus groups and prototype testing (conference room pilot) workshops.
  - o Outcome: Allows for greater engagement by meeting stakeholders where they are, instead of trying to get everyone into the same mode of communication. C-level will check out when presented with a communication channel that does not adhere to their level of engagement, and the mid/frontline will check out if they feel they are presented with too little detail to make effective decisions.
- Maintain open and transparent communication
  - o How?
    - Share progress, challenges, and changes consistently and openly
    - Acknowledge and address concerns openly and honestly. Do not allow turf-defense, as it will undermine the communication framework.

- Be careful that communications do not end up sounding like attorneys are protecting from litigation. It should be acceptable to say, "We missed this," without fear of retribution.
  - Ensure stakeholders understand how their input has been considered and what the outcome is, and by whom it was decided.
    - *Example*: Apply technical deep dives with IT executives, ROI discussions with finance executives, and hands-on demonstrations with those in Operations and also end-users.
    - Outcome: Creates real excitement by involving stakeholders appropriately and in line with their working styles and priorities. Creates an open level of communication where team members do not feel pressure to defend past decisions but can instead make changes and decisions based on a current understanding and new information presented throughout the engagement.

Effective Communication Frameworks

To surmount obstacles, organizations require formal communication frameworks that allow for collaboration without trespassing into departmental expertise. The following are established methods that offer specific outcomes when deployed:

1. The war room model

   The war room offers a high-intensity, physical or virtual, environment where interdisciplinary teams can tackle problems head-on. Communication, communication, and communication are the numbers one, two, and three on the list of how to keep a project on focus and make sure that you have a place where you can go where everybody can communicate.

   Implementation Guidelines:

- Create a regular rhythm (daily or weekly, based on project stage) of long-tail meetings where issues can be solved during the session.
- Have representatives with decision-making authority from every department.
- Prioritize meetings on particular concerns that need cross-departmental solutions.
- In meeting notes, record decisions and action items with explicit ownership.
- Keep project status, key decision points, and milestones visible to all involved.
- Clearly articulate where the engagement is in the journey.

The war room succeeds because it establishes a level playing field where department representatives concentrate on project success, not departmental interests. It provides dedicated space free from a structured meeting agenda where any topic of conversation, testing event, or information gathering can take place.

2. RACI framework

    The RACI matrix (responsible, accountable, consulted, informed) defines decision roles between departments (Matthews, 2024):

    R - Responsible: Who performs the work?

    A - Accountable: Who has ultimate decision-making power (just one individual)?

    C - Consulted: Whose input must be sought before decisions can be made?

    I - Informed: Whose progress and decisions need to be informed along the way?

    Implementation Guidelines:

    - Develop RACI matrices for important project processes and decisions.

- Confirm and discuss matrices with all stakeholders periodically.
- Apply the RACI when there are decision rights issues.
- Update the matrix throughout the project.

3. Visionary-integrator partnership

The visionary-integrator partnership is a leadership concept that distinguishes between two critical roles in high-performing organizations and sets boundaries and disciplines for how communication and decision-making should be structured between the two. The visionary sets the direction, dreams big, and generates bold ideas, while the integrator brings structure, translates vision into actionable plans, makes all key decisions, and ensures seamless day-to-day execution. This dynamic is essential for balancing creative ambition with operational discipline within a project.

The philosophy of separating the visionary and integrator roles was formalized and popularized by Gino Wickman in his book *Traction and Rocket Fuel* as a part of the Entrepreneurial Operating System (EOS)™, though its roots can be traced to historic business partnerships. One of the most celebrated examples is at The Walt Disney Company, where Walt Disney embodied the visionary, conceiving worlds, stories, and ambitions, while his brother Roy O. Disney assumed the role of integrator, deftly managing finances, operations, and the practical realities of building the Disney empire. This partnership was instrumental in Disney's early and lasting success, allowing both innovation and execution to flourish in tandem.

It is crucial to match visionary leadership with integrator implementation. The visionary is probably the most critical role to ensure the project goals and objectives are set, and the integrator is probably the most critical role in the project to ensure the goals established by the visionary are achieved. Understanding these roles and tailoring project communication to them is key to success. We will discuss this in detail in the next chapter.

4. Disagree and commit philosophy for decision-making and communication

    Intel's strategy, discussed later in this chapter in additional detail, offers a strong template for interdepartmental decision-making and a posture for how this can be implemented effectively.

    In general, this communication and decision-making framework allows teams to:

    - Create forums for open debate (disagree) and defined outcomes for implementation (commit)
    - Establish psychological safety for open departmental issues and even encourages disagreement while making clear the objective that all parties want the best outcome
    - Promote diverse thinking in planning and designing without initial pre-conceived decisions
    - Force all departments, once decisions are made, to fully support implementation
    - Remove the I-told-you-so syndrome by sharing success and embracing mistakes together

5. Celebration communication

    We have all seen status reports on a project where the email and attachment never get opened by the very individuals the communication was targeting. A frequently neglected element of status communication is the power of celebration, but celebration that works is more than pizza parties or artificial team-building activities. It is a matter of noting significant milestones, like the completion of process documentation, the successful testing cycles, the achievement of user adoption goals, or performance enhancement goals. During the celebration, recognition needs to be specific as well—spotlighting individual contributions, recording team accomplishments, publishing lessons learned, and broadcasting success stories.

Celebration communications should revolve around concrete benefits, such as proven time savings, quantified reductions in errors, enhanced customer satisfaction, and measured gains in efficiency. As another change manager explained it, "We began to celebrate small victories on a weekly basis. When the accounting group completed the documentation of their vendor management process, we didn't simply note in a status report that they had completed it. We had them present their work to leadership, spotlighted the progress they had made, and used their achievement to inspire other teams." With meaningful recognition and tangible results, teams can build motivation, engagement, and sustained progress.

An organization cannot expect consultants to have a perfect communication plan and team alignment in their briefcases or backpacks. Companies must invest in internal leadership specific to the transformation initiatives rather than feeble attempts at outsourcing this critical function. While vendors and consultants provide valuable expertise around communication or change management, ownership of this alignment must be internal. Some organizations achieve this by outsourcing a separate vendor solely to manage the project rather than having the implementation team perform both technical implementation and organizational alignment, but this should never become a way to skirt the responsibility of the project from within the organization or to create a break point between the organization and the project that can be replaced when things go poorly.

With effective communication, concrete outcomes, and committed internal ownership, teams can resurrect failing projects, generate new motivation, spark internal involvement, and create a culture of ongoing digital transformation that achieves the desired outcomes.

Case Study:
The CEO, the Test System, and the Mystery of the "Perfect" Financials

We stepped into the CEO's office, gripping a stack of SAP installation CDs. Yes, I said CDs. The mission is simple: update his PC's SAP client, verify functionality, and slip out before he asks us to help fix his email signature that has been broken since 2017.

The installation runs smoothly, as expected. Our process dictates that after every update, we log into the test company environment, just a quick check to confirm everything's working. Except this time, something is off.

He's already logged in.

A minor inconvenience, maybe. Until we realize something far worse.

He had been logged into the test system for a year or more since the last time we updated his SAP client during the annual update cycle.

Three hundred and sixty-five blissful, uninterrupted days of looking at the same sample data. There had been no logouts or resets of his system. Every time he opened the application, he'd seen only a perfectly preserved simulation.

At first, we exchange glances, suppressing laughter, unsure whether to break the news or quietly back away and let the illusion continue. Then, the real kicker hits us.

Days ago, in a high-stakes leadership meeting, this very CEO stood before his team and, with absolute conviction, declared:

"I log into SAP every day to review the financials and inventory numbers. It keeps me connected to the business."

Ah yes. Connected. To the business. The test business.

Year-old data.

A Case Study within a Case Study

Now, before we judge too harshly, this isn't exclusive to this CEO.

We once had a Sales Manager who said he actively used the CRM system. By "actively used," I mean his assistant would print out new leads, contacts, and opportunities from the CRM every week. These pages would be neatly organized into binders—literal, physical binders. His office was like a CRM museum exhibit: walls lined with binders labeled "Q3 Leads" or "Opportunities – 2019."

When asked why he didn't log into the CRM directly, he said, "I like to flip through the pages. It feels more real." Because nothing says "cutting-edge sales management" like flipping through outdated leads in a binder thicker than the CRM's user manual.

Now, back to the other story...

Suddenly, everything makes sense.

Why did the financials always look impeccable? Because they weren't current.

Why did inventory never fluctuate? Because nothing ever changed unless the database was changed to simulate a testing situation.

Why did he never spot trends or anomalies? Because there weren't any to spot.

For a year, he had been staring at ghost data, meticulously crafted by developers and testers, untouched by reality.

And yet, he spoke of these numbers with confidence, perhaps even pride, believing he had his finger on the pulse of the company.

We could say something. Expose the truth. But then again, the financials have never looked better.

The compounding problem with this is that the CFO had been pushing for a full ERP switch, blaming the company's outdated SAP system for its lack of real-time visibility. The stack of CDs would be the biggest data point showing this to be true. But there was a flawed assumption that a new modern ERP would magically provide instant sophistication to the organization.

The previous system had years of customization, structured data, and refined processes. Expecting a freshly implemented system to match that level of insight overnight was pure fantasy.

ERP replacements can cost millions and take years to stabilize. During that time, reporting often gets worse before it gets better due to historical data migration challenges. Even after go-live, customization, integration, and governance are still required.

A rushed ERP switch in 6-12 months just for the sake of modernization? A recipe for disaster. Business processes, integrations, and user adoption take years to refine. And when the new system still doesn't deliver the expected reports? Another system switch?

As we stood there, debating whether to break the news to the CFO or let the illusion live on, one thought lingered:

If the CEO spent a year confidently analyzing sample data ... would he even notice if the company made the switch away from the current dated system at all? Was he so disconnected from the reality of the organization that true digital transformation was only a mirage on the horizon? Or were the CEO and CFO so disconnected from the goals of the organization that they were both expecting a new system to solve for internal neglect and organizational dysfunction?

Open and complete communication is the key to establishing good requirements and producing lasting digital transformation results.

As an example of this, Intel's disagree and commit philosophy took shape in the 1980s, during Andy Grove's CEO tenure, when Intel was undergoing record growth and revolution in the tech sector (Intel Corporation, n.d.). Grove, having fled communist Hungary prior to contributing to the construction of one of the world's most powerful technology companies, knew that both vigorous debate and firm decision-making were critical to success in rapidly changing markets.

The philosophy was born of a practical need: how might Intel make decisions quickly enough to stay ahead while still having those decisions be informed by a variety of perspectives? The answer was a two-stage process that balanced careful disagreement with single-minded follow-through.

The philosophy is made up of two equal parts:

- First, "disagree" signifies that members of the team are not just permitted but meant to raise concerns, present contrary viewpoints, and challenge assumptions during the decision-making process. This creates an environment in which ideas are evaluated based on their merit and not on the basis of who proposes them.
- Second, "commit" means that once the decision is made after productive discussion, all individuals rally behind it regardless of their original position (Tunguz, 2017). The debate ends, and collective energy fully shifts to productive implementation. Grove codified this practice in his now classic management text *High Output Management,* and it was a pillar of Intel's operational excellence (Grove, 1995). The philosophy has since been used far beyond Intel, with variations being used at companies such as Amazon, where Jeff Bezos referenced it explicitly in his shareholder letters (Matthews, 2022).

Case Study: The Silent Room

The marketing department at a prominent consumer goods company should have been bustling with creativity and innovation. Instead, walking into their meeting room felt like entering a museum; all energy and ideas seemed to vanish the moment their Marketing Director arrived.

The conference room was set for their weekly campaign review. Glossy printouts of proposed designs were arranged neatly around the table, branded apparel filled the room, and coffee steamed in branded mugs. One by one, team members filed in, chatting animatedly about concepts and strategies.

Then she arrived.

The Marketing Director strode in with a commanding presence, her heels clicking authoritatively against the hardwood floor. The conversation immediately died. Notepads were straightened. Expressions neutralized.

"Let's not waste time," she announced, not looking up from her tablet. "The summer campaign concepts are disappointing. Again."

I had been brought in to help diagnose why the marketing department's system implementation projects consistently failed despite having several talented team members. They had implemented four different marketing automation tools in three years and still did not have the campaign-level attribution they needed. On paper, everything looked perfect. They had the right skills, adequate resources, and clear objectives. Yet execution repeatedly fell short.

The day before, I had attended a marketing planning session where the Marketing Director was not in attendance. She had been pulled into an emergency executive meeting. The difference was striking. That room had buzzed with energy, team members challenging assumptions, suggesting bold alternatives, building on each other's ideas. Their Associate Creative Director had presented a compelling counter-proposal to the current campaign direction, receiving enthusiastic support from colleagues.

But today, with the Director present, that same Associate sat silently, eyes fixed on her notepad.

"Rachel, you were leading the visual direction," the Director said, her tone sharp as she displayed a campaign mockup. "Explain why you thought this pastel palette would work for a product targeting young professionals?"

Rachel straightened. "I thought it aligned with the market research indicating—"

"It looks like a baby shower invitation," the Director interrupted with a dismissive wave. "Complete miss on the demographic."

No one spoke. No one made eye contact.

When asked about the alternative concept, the very one that had generated such excitement yesterday, Rachel simply said, "We've shelved that approach. Your original direction makes more sense."

This pattern repeated throughout the meeting. Ideas that had been vigorously championed in the Director's absence were abandoned without defense. Team members who had spoken confidently about market insights now deferred every question to the Director herself.

Later, in private interviews, the truth emerged.

"It's easier to stay quiet," one team member confided. "Last quarter, Jason challenged her campaign structure and spent the next three months assigned to updating spreadsheets."

Another added, "We learned it's more efficient to wait until she's not around to have real discussions. Then we figure out how to package our ideas so they seem like they were hers to begin with the next time we meet with her."

The consequences were severe. The department had developed a shadow operating system, appearing collaborative while practicing what amounted to organizational theater. Senior leadership couldn't understand why systems and campaigns consistently underperformed despite apparent consensus in planning meetings. The marketing team was developing a reputation for overpromising and underdelivering.

Most troubling was the invisible toll: innovative ideas never surfaced, critical flaws went unaddressed, and talented team members were actively interviewing for jobs elsewhere.

This environment was the antithesis of Intel's disagree and commit philosophy. The team had created an illusion of immediate agreement that masked deep misalignment and passive resistance.

Without psychological safety to voice concerns, team members resorted to malicious compliance, nominally agreeing while knowing plans would fail. The appearance of perfect alignment had actually created perfect dysfunction.

The solution began with a candid conversation with the Marketing Director about the shadow system that had developed. While initially defensive, seeing concrete evidence of how performance suffered, and learning that two top performers were planning to leave, created a moment of recognition.

Over the following quarters, we instituted structured feedback mechanisms where dissenting views were not just permitted but required. Each major decision needed at least two alternative approaches to be considered. Anonymous feedback channels gave team members safe ways to express concerns. Most importantly, the Director began acknowledging when her initial directions were improved by team debate.

The transformation wasn't immediate, but nine months later, the change was palpable. In a campaign planning session, I watched as that same Associate Creative Director confidently challenged the Director's approach, and the Director not only listened but ultimately adopted many of the suggested changes.

The impact extended beyond improved team dynamics to real technical outcomes. Their recent marketing automation platform installation finally produced the campaign-level attribution for which they'd yearned for years.

Previously, when email metrics loudly declared dismal click-through rates drastically lower than industry averages, no one dared contradict the Director's template designs or segmentation strategy. Instead, the technology was criticized by team members, who built complex manual workarounds in spreadsheets that made the numbers appear better but offered no real insight or repeatability. When the platform attribution revealed that some campaigns were performing far below expectations, team members intuitively knew why but would not say anything. Customer segments were not accurate, email content wasn't tied to landing pages correctly, and call-to-action buttons were not functional—all technical issues that weren't addressed because no one felt safe in criticizing the Director's approach. Instead, the system took the blame because it could not fight back.

With psychological safety present, these technical issues were finally addressed openly. Three months later, email engagement metrics doubled, the manual attribution spreadsheets were put out to pasture, and for the first time, the team could effectively measure which campaigns were generating actual revenue versus activity metrics.

"This is way better than what I proposed," she acknowledged to the room. "This is why we need everyone's perspective."

The lesson in this case study is that if there is no disagreement, you haven't built a leadership team; you've assembled discontented lemmings who are quietly dying inside. True collaboration requires psychological safety, structures that encourage productive conflict, and leaders secure enough to have their ideas challenged. Without these elements, alignment becomes an empty theatrical performance rather than a foundation for success.

With this goal in mind, here are some ways that Intel's philosophy fosters a healthy organization:

Way #1: Encourages Open Dialogue

The "disagree" component creates several vital benefits for organizations:

It first establishes psychological safety, the mutual belief that team members will not be reprimanded or embarrassed for contributing ideas, questions, concerns, or mistakes. This safety is the cornerstone of open communication and is repeatedly found by researchers to be a critical predictor of team effectiveness.

Second, it offers decision-makers a variety of viewpoints. Having this variety of thought detects issues earlier, reveals possibilities that would not otherwise be looked at, and results in more concrete solutions.

Third, it avoids the "emperor's new clothes" phenomenon, where obvious weaknesses are not discussed because nobody thinks they can speak out. Where disagreement is not tolerated, critical thinking dies and groupthink rules.

Finally, it is an early warning system for flawed approaches. Since team members can freely speak up, organizations reap the advantage of their collective wisdom and experience before making significant investments in potentially erroneous directions.

Way #2: Fosters Commitment

The "commit" portion of the philosophy is equally crucial for organizational effectiveness:

It generates coherent execution post-decision points. When discussion concludes, everybody's energy is focused on effective implementation instead of continuing to debate the decision or passively resisting the direction selected.

It avoids the I-told-you-so syndrome when team members who were against an initiative take pleasure in its failure instead of putting their best foot forward to make it a success. Instead, even those who had disagreed earlier are likely to utilize their entire potential to make the selected approach function.

It speeds up implementation by removing recurring debates that consume energy and cause confusion. It enables teams to move forward with confidence instead of constantly revisiting matters already resolved.

It generates collective ownership of results. As everybody buys into decisions, failure or success becomes collective and is not blamed on the initial proponents.

Way #3: Reduces Decision Paralysis

Many organizations suffer from analysis paralysis, an inability to act because of endless debate, fear of making the wrong choices, or a lack of clarity about decision rights. The disagree and commit approach is a straightforward antidote to this:

It gives a clear cutoff for debate and a launch pad for action. Teams understand when they are in the "disagree" phase and when they are in the "commit" phase, creating natural boundaries for argument.

It establishes decision ownership, in which leaders make final decisions after taking input rather than trying to reach everyone's consensus (which more often than not results in diluted compromises that frustrate all).

It accepts that trade-offs and uncertainty will accompany most business decisions and that perfection is improbable. Instead, it focuses on decisions made soundly and implemented effectively.

Way #4: Builds Trust

Trust forms the foundation of effective teams, and disagree and commit nurtures trust in several ways:

It demonstrates respect for team members' expertise and perspectives by actively soliciting their input before decisions are made.

It creates transparency around decision rationales, helping people understand why certain directions were chosen even if they initially advocated for alternatives.

It builds confidence that concerns will receive fair consideration rather than being dismissed or penalized, encouraging continued engagement.

It creates predictable decision processes where team members understand how and when their input will be incorporated, reducing anxiety about how decisions are made.

Way #5: Enhances Accountability

The philosophy creates a framework for meaningful accountability:

It establishes clear expectations that everyone will support decisions once made, making it easier to identify when team members aren't fully committed.

It creates shared responsibility for outcomes since everyone had the opportunity to influence the direction and commit to supporting it.

It eliminates common excuses for poor performance like "I never thought this would work" or "This wasn't my idea," focusing instead on collective commitment to success.

It clarifies the distinction between healthy debate (during decision-making) and unhealthy resistance (after decisions are made), setting clear behavioral expectations.

Way #6: The Psychological Aspect

The disagree and commit approach aligns with fundamental principles of human psychology:

It meets the human desire for voice and closure; individuals need to be able to be heard, but also require clear direction instead of continued uncertainty.

It generates psychological ownership since it engages individuals in decision-making and intrinsically motivates them to ensure project success.

It reconciles conflicting psychological needs for both autonomy (having control) and reassurance (having direction).

It recognizes that commitment follows involvement and people invest in what they help shape, so the disagree phase is necessary to generate follow-through commitment.

Intel's Application in Practice for Digital Transformation

Using disagree and commit when implementing digital transformation initiatives sets a strong foundation for success:

During system choice, foster open discussion regarding platform weaknesses and strengths, and then commit wholeheartedly to gaining maximum value from the chosen system instead of lamenting or revisiting the decision.

At design phases, foster diverse input on workflow and process solutions, then fully support chosen designs to prevent revisiting decided design mid-development.

For technical architecture choices, guarantee thorough discussion of the alternatives, followed by collective agreement on chosen approaches without continuously changing decisions.

During implementation planning, encourage constructive discussion regarding order and priorities, and then commit together to the agreed map instead of continuously proposing different modes of approach or new priorities.

Intel's disagree and commit philosophy provides a potent means to achieve a balance between considered deliberation and committed action. By fostering settings in which ideas can be debated prior to decisions and then acted upon completely after decisions, organizations not only develop improved solutions but also more reliable execution. In the intricate realm of digital transformation, where innovation along with alignment takes precedence, this balanced strategy of communication delineates a well-tested roadmap for success.

The Foundation-First Approach: Have we created a foundation for a reset of expectations?

When a project is off the rails and in need of a reset, organizations need to address the following checklist of fundamental factors before they should think about a large-scale digital transformation or implementing new systems. The question should be asked: "Are we even engaged in the right initiative?" A reset is the perfect time to put this question on the decision-making table. A new system implementation or replacement may not always be the right answer. Ask your organization if these key components are in place prior to evaluating a change in system:

✓ Assess data structure and governance

Why it matters:

Poor data structures and poor data governance are the root causes of visibility issues. No system, no matter how expensive or feature-rich, can generate good reporting from bad or incomplete data.

Main issues to consider:

- Where does your critical business information reside today?
- What is the quality of this data (completeness, accuracy, timeliness)?
- Do you have standardized data models and data entry procedures?
- To whom does data quality belong for each key data domain?

Impact of getting this right:

When correctly set up, data governance gives organizations a solid grasp of their information assets. It is essential, whether you introduce new or keep current systems.

✓ Invest in a data warehouse or control tower

Why it matters:

Most companies today don't run on a single system. Having visibility into information requires tying data together across systems.

Main issues to consider:

- Data warehouse or data lake: Unify data from systems in a structured environment optimally used for reporting.
- Control tower solutions: Offer end-to-end visibility across operations without ripping and replacing core systems.
- Integration platforms: Link current systems to exchange information more efficiently.

Impact of getting this right:

Organizations achieve cross-system insight without disruption and the expense of complete system replacement.

✓ Leverage business intelligence and reporting tools

Why it matters:

Most companies utilize less than 50% of their current system's functionality. There is usually hidden potential in your current environment.

Main issues to consider:

- Data audits: Determine the hidden insights in your existing system's data.
- Module activation: Activate unused modules that can bridge existing gaps.
- BI customization: Create custom analytics or measures across the current systems.
- User training: Ensure users understand available tools and how to answer questions for themselves without having to rely on software engineers to build new tools.

Impact of getting this right:

Significant system value, achieved many times very quickly, and performance enhancement without the disruption of replacement.

✓ Optimize the current system's reporting capabilities

Why it matters:

Most organizations use less than 10% of their existing system's out-of-the-box reports. There is typically untapped potential in your current reporting ability.

Main issues to consider:

- Configuration audits: Determine the unused reports in your existing system.
- Report activation: Activate unused reports that might bridge existing data visibility gaps.
- Reporting customization: Create custom reports within the current framework.

- User training: Inform users about available reporting tools and how to build their own self-service data tools or reporting capabilities.

Impact of getting this right:

Substantial system data visibility improvement without the disruption of replacement.

✓ Only consider an ERP or system change when there's a structural need

There are many times when you will need to replace systems, but data visibility should not be the primary driver.

- If the current system truly cannot support business operations, then a switch might be necessary.
- If you have selected a system that does not provide technical access to data, then you may have to implement solutions that provide visibility or replace the underlying systems.
- If the current systems contain security vulnerabilities or severe technical limitations due to age or design decisions, then a replacement could be the correct option.

System replacement should not be the initial choice for visibility or reporting issues. It is costly, disruptive, and does not necessarily heal the underlying causes.

The wiser course is to establish a solid base in the form of good data governance, analytical capability, and process standardization. With this base in place, organizations are in a position to make more informed decisions on system needs and typically discover that their current systems, when adequately configured and augmented with the appropriate analytics tools, can meet their requirements.

Keep in mind that data visibility issues are not generally addressed by merely altering or changing the system. They need a holistic approach to how data is established, sustained, and leveraged throughout the firm. This could be a very different project from the one you are currently engaged in.

Reassessing and Simplifying

When projects begin to fail due to poor requirements-gathering and design, the first step is honest assessment. Consider this checklist:

- ✓ What specific issues are users experiencing?
- ✓ Where are workarounds being created?
- ✓ Which processes are breaking down?
- ✓ How severe is the impact on business operations?

Document everything, no matter how minor it seems. Often, small issues point to larger systemic problems. Then, review every feature and requirement against three criteria:

- Is it essential for business operations?
- Does it address a critical pain point?
- Can it be implemented without major customization?

Be ruthless in eliminating nice-to-have features. Remember: complexity is the enemy of success. So, to simplify the process, look for opportunities to:

- Remove unnecessary approval steps
- Standardize common procedures
- Eliminate redundant data entry
- Consolidate similar processes

The goal is to make processes as straightforward as possible before applying technology. Assess each technical requirement against:

- Business value
- Implementation complexity
- Resource requirements
- Maintenance needs

Focus on configuration over customization whenever possible.

Rebuilding Confidence and Momentum

Recovery isn't just about fixing technical issues; it's about rebuilding team confidence and trust when projects begin to fail because of the people and processes involved.

This requires:

1. Quick wins
   - Identify immediate improvements
   - Demonstrate visible progress
   - Celebrate small successes
   - Share positive outcomes

2. Team rebuilding
   - Reset roles and responsibilities
   - Provide needed training
   - Establish clear communication
   - Foster collaborative problem-solving

3. Progress tracking
   - Set measurable milestones
   - Honestly track and share progress
   - Address issues quickly
   - Maintain full transparency

Moving Forward

Successful digital transformation requires more than just selecting the right technology. It demands a clear understanding of current processes, realistic assessment of organizational readiness, and a structured approach to managing complexity. To do this you must have strong team alignment and engagement, and a clear recovery strategy for when things go wrong.

Most importantly, it requires the wisdom to know that technology alone cannot fix broken processes or unclear objectives. True transformation starts with people and processes; technology simply enables change. Reengaging in the foundation is merely the initial step of an effective change. Robust processes, structures, and recovery plans are all crucial, yet they are worthless unless the stakeholders are also aligned with the overall objectives.

Reflection Questions

1. In your current or recent projects, what signals indicated the need to rebuild the foundation? Were they technical issues or people/process problems?
2. Think about a time when automation failed to deliver expected benefits. Looking back, what foundational elements were missing?
3. How does your organization currently assess process readiness? Are you measuring the right things?
4. When projects start to fail, does your team focus more on technical fixes or foundational issues? Why?
5. Has your organization engaged in finger-pointing, blaming the vendor or the technology when things begin to slip? Why?
6. What specific steps could you take starting tomorrow to start rebuilding the foundation of a struggling project?

CHAPTER 4

The Project Bill of Rights: Establishing Psychological Safety and Clear Expectations

"The single biggest problem in communication is the illusion that it has taken place."
—George Bernard Shaw

The conference room fell silent. The project manager had just delivered the newest change in requirements from the executive group, the fifth significant change in six weeks. Developers looked knowingly at each other across the table, business analysts gazed into their notebooks, and the technical lead's jaw clenched noticeably.

"I know what you're all thinking," the project manager said, "but we must be flexible. The department leads say this is a must-have requirement."

A junior engineer hesitantly held up her hand. "But we've just completed re-building the inventory module from the last set of changes. This will put us back at least three weeks, and we're already behind schedule."

This is something that happens in companies everywhere, every day. Projects fail at these critical junctures because of the absence of psychological safety and the freedom within the team to speak openly about problems or potential changes.

In that conference room, each person experienced it in his or her own way, albeit in similarly destructive manners. The developers were frustrated and defeated, realizing they would have to bear the burden of unrealistic timeline expectations. The business analysts were torn between business requirements of the department leads and technical limitations expressed by the engineers, unable to balance the two. The technical lead was frustrated by the coming decision to compromise quality for the sake of a timeline or forced to have the team work extreme hours to add the new requirements into the existing timeline. The project manager was caught in the middle between executive pressure and team capacity. People in all roles felt the pressure and frustration mounting within an already strained project.

Psychological safety, a phrase coined by Harvard professor Amy Edmondson, is the confidence that someone can say what they think without punishment or embarrassment. It's mutual trust that the team won't embarrass, reject, or punish anyone for speaking up with ideas, questions, concerns, or mistakes. When that safety is compromised, as it had been in that conference room, projects start their inevitable downward spiral. Team members don't speak up early when problems are small and can be fixed. Creativity disappears as people cling to secure, conventional solutions. And ultimately, the organization forfeits its ability to learn and adapt, the very capabilities most required during complex implementations.

"That's not your problem," was the curt response. "You need to get it done."

The temperature in the room felt like it had fallen by ten degrees. Something essential snapped that day in the mentality and trust within the team. Six months later, the project would be canceled after it had consumed millions from the budget with nothing to deliver.

This is a situation that unfolds in companies all over the world every single day. If members are afraid to share concerns and openly discuss what they believe is true, without fear of retribution, then failure has already been planted.

The antidote to this toxic dynamic is what I like to call a Project's Bill of Rights, a foundational set of decisions that establishes clear expectations, generates psychological safety, and empowers all stakeholders to authentically contribute to project success.

Why Project Teams Require a Bill of Rights

In the U.S., the Bill of Rights was enacted in 1791 as the first ten amendments to the Constitution, written mainly by James Madison. This seminal document guarantees fundamental liberties like freedom of speech and religion and imposes absolute constraints on governmental power. Its brilliance lies in the realization that even benevolent power needs to be limited lest it be exploited, and that some individual rights need to be safeguarded against the majority. These values have supported American society and law for more than two centuries.

Similarly, a Project Bill of Rights protects the fundamental needs for teams to thrive and places constraints on how stakeholders, leaders, and team members engage with each other.

Just as constitutional rights protect citizens from governmental overreach, these project rights protect team members from the natural tendency of organizations to prioritize short-term objectives over team welfare, realistic constraints, and long-term success. And just like constitutional rights, they establish a shared expectation that certain principles should remain unbroken, even under pressure.

We can see the need for such a system when we examine some core reasons why projects fail:

- Fear to speak: Team members recognize essential problems but remain silent for fear of retaliation. This generates anxiety and disengagement as individuals feel complicit in a failed strategy.
- Scope creep: Changes add up without timeline or resource adjustments. Members burn out and feel resentment. Quality suffers as they attempt to cram growing requirements into rigid parameters.

- Blame culture: Failures become opportunities for finger-pointing rather than learning and evolving. This evokes defensive behaviors, documentation hoarding for self-protection, and avoidance of risk-taking that can lead to a lack of innovation and creativity.
- Burnout: Teams operate at unreasonable levels, compromising quality and well-being. Burnout produces more mistakes, loss of creativity, and ultimately cynicism regarding the very organizational values they hope to uphold by investing in the project.
- Lack of transparency: Problems are concealed until they are catastrophic. This generates mistrust, results in an information vacuum filled with rumors, and bars early course correction when difficulties are still manageable.
- Unrealistic expectations: Teams take on unrealistic deadlines instead of pushing back. This fosters a culture of impossible commitments, which results in normalized deception where green status reports hide major issues.

Take, for example, a pharmaceutical company that was introducing a new supply chain management system. Halfway through the six-month project, it became apparent that integration with legacy warehouse systems would require double the time originally anticipated.

The technical lead was aware but was not willing to speak up, having witnessed a colleague being taken off an earlier project for bringing the same bad news. So, he told the team instead that they were "slightly behind but catching up." By the time the full magnitude of the delay was understood, the project was irretrievably behind schedule, stakeholder confidence had disappeared, and the company lost the opportunity to course correct with all of the information.

A Project Bill of Rights would have averted this catastrophe by allowing for open communication, setting realistic expectations for scope changes, and providing a work environment in which problems could be fixed before they became disastrous or priorities could be shifted to keep the timeline intact.

The success of this strategy, however, depends on its acceptance by all parties. Like a social contract, these rights only work when everyone, from executive sponsors to team members, recognize and respect them.

If even one of the key stakeholders, particularly someone in authority, ignores or violates these rights, the entire system begins to deteriorate. Team members quickly spot the gap between espoused values and action, at which point psychological safety evaporates.

This is why the development of a Project Bill of Rights needs to start with a sincere commitment at the highest leadership levels, with outward compliance, particularly when it is most difficult to do so.

Developing Your Project Bill of Rights

A comprehensive Project Bill of Rights will establish clear expectations in several key areas. Let's discuss each right in detail:

1. The Right to Disagreement

"We have a right to disagree constructively without fear of reprisal."

Disagreement is not betrayal of another's idea; it often is the genesis of creativity and risk mitigation. When team members feel safe to disagree, the project benefits from multiple opinions and is protected against groupthink.

Groupthink, a term created by psychologist Irving Janis in 1972, happens when the need for conformity or harmony in a group leads to dysfunctional decision-making. It is human nature to want to fit into a group, which is why we need to strive to form groups that are built around disagreement.

Janis's studies of foreign policy blunders, such as the Bay of Pigs invasion, indicated how catastrophic decisions can be made by geniuses if there is no space for dissent. The Bay of Pigs invasion exposed flaws in planning, poor communication, and unrealistic expectations among U.S. leadership and their Cuban counterparts, resulting in a disastrous military operation. More recent research by Charlan Nemeth at UC Berkeley has shown that leadership groups that are subjected to dissenting viewpoints deliberate on more alternatives and make improved decisions, even if the dissenting view itself proves to be wrong (Janies, 1972).

Without constructive conflict, groups fall prey to confirmation bias, premature solution design, and collective overconfidence, all deadly poisons for complex technology projects. However, building a team entirely devoid of conflict-avoidant members is an endeavor fraught with complexity. Human nature, shaped by the instinct to preserve harmony, often nudges individuals toward sidestepping confrontation, even when such avoidance undermines the very foundation of progress. The team must intentionally nurture an environment so robust in trust and openness that every participant, regardless of personality or past experience, feels both empowered and compelled to voice dissent. Even then, it is rare to see this ideal take shape in perfect balance because the fear of retribution, isolation, or disrupting group cohesion has deep roots. A team where no member flinches from constructive conflict, yet can align mentally and emotionally with any conflict's outcome, is a feat as demanding as it can be transformative.

Implementation:

- Recruit the team with the intentionality of dissent and alignment.
- Create structured forums for questioning that demand voicing various opinions.
- As role models, leaders accept and react positively to criticism.
- Separate criticism of an idea from the person proposing it.
- Document and act on all matters brought up, even if the initial action is taken.

The price of silence about conflict can be catastrophic. Consider Air Florida Flight 90, which crashed into the 14th Street Bridge in Washington, DC in 1982 and killed 78 people. It was discovered through investigation that the copilot had expressed concern about the buildup of ice on the wings multiple times during takeoff. The captain shrugged off such concerns repeatedly, generating a dynamic whereby warnings from the copilot grew more tentative. Eventually, the NTSB determined that the crash would have been avoided had the captain taken these warnings seriously (Robinson, 2025). In project environments, the stakes are not as high, yet the dynamic remains the same. When team members are not safe to speak out forcefully, or when leaders brush aside those warnings, catastrophe is not just a possibility but a certainty.

Example in Practice:

One practical way to nurture psychological safety and counteract the fear of dissent is to invite the team to participate in a "Ten Bad Ideas" exercise. In this activity, the group is tasked with deliberately generating a list of ten purposefully bad, impractical, or even humorous solutions to the problem at hand. By lowering the stakes for contribution and making it clear that outlandish or "wrong" ideas are not just tolerated but actively encouraged, teams disrupt the collective impulse toward only sharing safe, consensus-driven suggestions. This lighthearted approach sparks laughter and camaraderie, but it also unlocks new perspectives by loosening mental constraints. The seeds of breakthrough solutions are often hidden amid the "bad" ideas. Moreover, this exercise signals that every voice is welcome, and that the process of voicing dissent, critique, or unconventional thinking is a valued ingredient in the group's collective wisdom.

2. The Right to Miss a Requirement

"The team has the right to acknowledge that a requirement was missed and to accept scope modifications with a corresponding change to (cost) timeline and/or resources."

Scope changes are unavoidable within any technology project of size, but scope changes with fixed resources and fixed timelines are a disaster waiting to happen. This right guarantees that scope, time, and costs are intertwined. When you modify one, you must revise the others accordingly.

This sounds like common sense, but it is amazing how many project teams simply accept new scope without forcing the issue of timeline and costs. The illusion that a change in one lever does not impact the other two is a fallacy that is all too often accepted by the team members or consultants involved in the project. This right forces the issue that there must always be a change to cost or timeline when scope changes and provides an openness for anyone involved to say, "We missed a requirement," without fear of retribution.

For this right to be effective, two underlying elements are necessary. First, the project team should have a clear, documented, and approved scope baseline to which all stakeholders have formally committed. This level is what we are calling Level 3-5 requirements. In the absence of this shared point of reference, attempts at negotiation of changes are relative and many times imaginary. Second, there should be a defined process for establishing and classifying what a change is as opposed to a clarification or refinement of existing scope. This is only possible when story level requirements, as well as functional and technical requirements, are documented and approved by the appropriate team members.

The team must assign roles to people who possess the ability to approve or reject scope alterations. Under normal circumstances, workstream leaders ought to possess approval authority for changes in their area, and a steering committee or project sponsor should approve significant changes. If this hierarchy is not present, then scope management turns into a political undertaking as opposed to a formal procedure.

Implementation:

Perform a thorough impact analysis for every proposed change, analyzing:

- Schedule impact: How many days/weeks would this add to the schedule?
- Resource impact: What other personnel, equipment, or software would be necessary?
- Technical complexity: How would this affect system architecture, integration points, or performance?
- Risk profile: What other requirements would be at risk as a result?
- Opportunity cost: What planned work would need to be ranked lower in order to accommodate this change?

Example in Practice:

During the final stages of user acceptance testing (UAT), a missing requirement surfaces, one that was supposed to be obvious, yet managed to slip through every review and checkpoint. This situation tends to shift the atmosphere in the project quickly if allowed. Team members will begin scouring emails, requirements documents, and meeting notes, hoping to find evidence that someone has dropped the ball. Fingers will be pointed: Was it the business analyst who failed to document it? The developer who presumed it was out of scope? The tester who hadn't flagged the gap earlier? The uncomfortable truth is—everyone missed it.

Even with meticulous analyses, documentation, approval processes, and sign-offs, something will fall between the cracks in every project. This fact should be both accepted and acknowledged. Consulting firms in particular have a tendency to start guarding what they say in these situations, because of fear of legal retribution, impact to the budget, or that the client will request credits. In turn, people start acting like attorneys instead of team members.

But with the team's own Project Bill of Rights, they knew the next steps weren't about denial, accusation, or endless debate. When a missing requirement is discovered, the budget and timeline must be adjusted accordingly, or something lower in priority must make room. Instead of scrambling to assign fault, the team can open the change impact tracker and perform a thorough analysis: How will this requirement affect the schedule? Which resources would it demand? What other features would need to be deferred or dropped to accommodate it?

With the facts in hand, they can bring the situation to the team which will review the objective data. Together, they agree to shift a minor feature to a later release, adjust the timeline, and communicate the change transparently to all stakeholders.

In the end, what could have been a divisive setback became a testament to their commitment to open, honest scope management, where the focus was less on blame and more on building a solution everyone could stand behind.

One practical way to nurture this behavior is to resist setting a go-live or launch date of a new system or feature until UAT has begun and there is a validation of the requirements. Resist the demands to establish a finish date before the requirements have even been gathered. Also, set a budget for requirements-gathering separate from the budget for implementation, UAT, training, and go-live. Having separate budgets and budgeting in sequence resists the temptation to set a budget for the engagement prior to completion of detailed requirements-gathering.

3. The Right to Honest Prioritization

"We have a right to speak our candid assessment of priorities, even if it places us in opposition."

Not all requirements are created equal, and making them equal wastes time and misses opportunities for real, impactful change. Teams need to be able to question priorities and provide alternatives from their perspective. Features range widely in business value, cost of implementation, risk, and alignment to strategy. A particular feature may be a must-have for one department, yet add minimal overall value to the organization. A single feature can have a clear technical implementation, while a supposedly simple request by another can encompass complex architectural revisions. In the absence of objective criteria for determining priorities, organizations will inevitably turn to political ones; the loudest department or one with the largest organizational clout gets its features done first.

Cross-departmental priority conflicts are especially challenging because every department prioritizes according to its own departmental goals. Healthy organizations solve these conflicts by creating a prioritization process that emphasizes organizational priorities over departmental priorities. This typically includes a steering committee with representation from all the stakeholder departments, explicit decision criteria linked directly to strategic objectives, and an escalation process for prioritization conflicts.

Implementation:

- Develop objective ranking criteria based on the objectives of the project.
- Create a shared priority matrix available to all stakeholders with simple 1-5 rankings.
- Hold regular priority review meetings and be open to shuffling priorities.
- Include a revenue impact, expenses impact, or time impact to the priorities so it can be clearly articulated to the decision makers what the business impact of the requirement is.

Example in Practice:

When one insurer was building a new claims handling system, their IT director presented a simple 2x2 priority matrix that plotted features by business value (low to high) and implementation complexity (low to high). In a review session, the IT team illustrated that numerous high-complexity, low-value features were absorbing disproportionate effort. Instead of provoking conflict, this objective analysis enabled the team to redirect effort to high-value, low-complexity features that provided immediate business value.

One strategy is to play priority poker and have people vote on the quadrants for each item with cards or an online vote to make the voting anonymous. Stakeholders can therefore anonymously vote on where features fit in a prioritization matrix. This prevents corporate influence bias, where some staff simply repeat what they believe senior leaders want. When votes reveal significant divergence in how varying stakeholders assess priorities, it allows for constructive debate of assumptions and perspectives.

4. The Right to Fail

"We have the right to fail at something and try it again without shame or blame."

Innovation needs experimentation and experimentation comes with failure. If failure is positioned as a chance to learn instead of a chance to find someone to blame, then teams can iterate their way to great solutions.

Any real digital transformation project needs to incorporate some aspects that are outside the current skill and capacity of the organization. If it doesn't include those groundbreaking aspects—new tech, new processes, or new approaches—then the project isn't transformative.

This transformation implies that every major initiative must have proof-of-concept, pilot, or controlled tests in which the team can test hypotheses and learn through small, contained failures before deploying at scale. Punishing the inevitable errors that accompany this discovery doesn't make the project safer; it ensures that teams will adhere to tried-and-true, safe processes that constrain the true value of the transformation.

Implementation:

- Hold blameless postmortems following failures. Discuss the test case, hypothesis, and result.

- Celebrate the positive, productive failures that provide valuable insights.
- Add pilot and proof-of-concept time to project schedules.
- Record lessons learned and integrate them into future project activities or use them to finalize design decisions between L3 and L5 requirements.

Example in Practice:

One financial services firm was introducing AI-driven customer service reps to deal with standard inquiries. In early trials, the team found the AI was failing to adequately comprehend close to 40% of questions from customers regarding loan applications. The project leader had established a learning mindset from the beginning, and it was made very clear that initial failures were anticipated and beneficial.

Rather than dismissing these terrible results or blaming the AI vendor, the team undertook an in-depth study. They found that customers were utilizing industry jargon and acronyms that were not present in the AI's model training data. They also found that the company's own loan products had naming conventions that perplexed the AI's natural language comprehension because they had multiple industry meanings.

This initial failure served as the impetus for developing an exhaustive financial services terminology database, renaming customer-facing products to be more unique, and establishing an ongoing process for maintaining and improving the AI's language models.

"That first failure of the tests was actually our greatest success," the project manager later said. "If we'd kept pushing forward, thinking that things would improve on their own, we'd have launched an AI that would have frustrated customers and damaged our reputation. Instead, we have an AI assistant that gets it right 93% of the time now, better than some of our human representatives."

One practical way to encourage this behavior is to add a formal conference room pilot phase between the requirements-gathering phase and formal implementation. This series of meetings is where users gather in a physical or virtual conference room to act out the various requirements using real-life examples in the system. The next step is to formally document what is believed to be true, what should be proved, and the actual outcome.

5. The Right to Change a Requirement

"We have the right to revise requirements as we get more information."

Defining requirements is a process, not a moment in time. As teams implement, see the system in action, identify edge cases, or even just progress in their understanding of the system, they always find gaps, inconsistencies, and detail needed in the initial requirements.

Requirements inevitably need to be revised for valid reasons. Business contexts change during implementation, making some original requirements redundant and others new. Subject matter experts cannot recall every edge case or exception in intricate business processes. Team insight is refined as they get deeper into the system functions and understand how it operates and how it handles situations. And occasionally, requirements that were seemingly obvious in theory only manifest their ambiguity when implementation begins.

This is not a failure of individuals, or even a lack of attention to detail by any individual, but a reality of complex projects. No single person or document can get every requirement in every area perfectly right the first time. By acknowledging this reality and creating a blame-free process for revisiting requirements, we shift from "who dropped this?" to "how do we handle this now that we've discovered it?" The team becomes collectively responsible for requirements evolution, acting as each other's safety net rather than blaming when gaps emerge.

Implementation:

- Perform early education on the new technology for all parties involved to provide a better understanding of how a new system may handle core requirements.
- Hold periodic requirements review meetings to assess what has been discovered.
- Keep a living requirements document that develops with the project. Only estimate once they are believed to be final and reassess when things change.
- Define a clear process for recording new requirements.
- Schedule stakeholder time for requirements refinement and review sessions.
- Create a formal process to rank newly found requirements against their current priorities in the backlog so that they compete on a level playing field for resources and aren't automatically pushed to the top.

Example in Practice:

A financial services company rolling out a CRM system set up a bi-weekly requirements refinement meeting in which the team shared new things learned during implementation. In one of those meetings, they discovered that regulatory requirements around customer data had shifted since the original specifications had been written. Instead of forging ahead with out-of-date requirements, they were able to integrate the new regulations without halting the project, because the team had clear permission to go back and revisit requirements.

6. The Right to Ask Questions

"We have a right to ask any question without being seen as incapable."

Discovery questions form the essential bridge between technology experts and process owners, serving as the means by which each side can illuminate the mysteries of the other's domain. Those fluent in new systems rely on questions to unravel the nuances of existing workflows, while those versed in business processes must seek to understand the capabilities and limitations of new technologies. Team members must possess this curiosity, and it should be encouraged.

One of the biggest barriers to effective communication in projects is the use of acronyms and technical jargon that are not defined. Organizations even create their own vocabulary over the years, as well as acronyms with various meanings in various departments. Too frequently, people in meetings sit quietly instead of asking for clarification for fear of sounding stupid. In fact, every time a person says, "What does ASMG stand for?" there is nearly always another person in the room grateful someone asked the "dumb" question they were also going to ask. I have been in many meetings where it's discovered that within the same department, an acronym is not understood by the team, but many people were afraid to ask the question for fear of being the only one in the dark. There are also instances where the same TPS acronym has one meaning to the technology team and a completely different meaning to the department that uses the tech.

Engaging a consulting firm to conduct a detailed discovery and requirements-gathering process is a great way to uncover many of these discrepancies of vernacular within the organization. They come in with little to no knowledge of the organization and begin asking a series of structured questions. They can be the ones ignorant of the process, asking the "dumb" questions for the benefit of everyone involved.

Implementation:

- Leaders need to set an example by asking questions, even questions they know the answer to but suspect others may not, opening the door for everyone.
- Conduct special business terminology sessions in which the express purpose is clarification.
- Maintain a project glossary of terms, acronyms, and jargon that is accessible to all team members and updated regularly.
- Recognize and reward clarifying questions that prevent misunderstandings.

Example in Practice:

One tech firm implementing supply chain software put up an online question board on which members were able to ask questions anonymously about the project. In weekly meetings, the project manager answered the questions without saying who had asked them. By doing so, the significant misconceptions regarding requirements that would not have been otherwise seen were brought to light since team members were not afraid of looking dumb.

One practical example is to start off every new project with a word cloud exercise where a whiteboard or posterboard is used within a meeting room or space that will be used throughout the project. Have a designated spot where project team members can write acronyms, jargon, technical words, or phrases that need to be defined. This can be a fun exercise when, throughout every meeting, various acronyms are written on the board as they are used by members of the meeting.

7. The Right to Challenge Deadlines

"We are entitled to question all deadlines or resource limitations at any time."

Arbitrary deadlines and inadequate resources condemn projects to failure. It has been amazing to me to see the number of first discussions with clients where they ask the question, "How long will it take before we are ready to go-live on the new system?" We have not even gathered a single hard requirement, and the executive team wants a date on the calendar to shoot for. While this seems like goal setting 101—establish a goal and a target date to reach it—in complex technology engagements, this is fool's gold waiting to be sold by any vendor willing to give the executive team the answers they want, knowing full well that there is not yet enough information to answer the question with any level of intelligence. Teams need the authority to resist constraints that compromise quality, given too early in the process, or ones that make success unrealistic.

Leaders do have the authority to override objections and order "do it anyway," founded on business reasons that might not be clear to everyone in the team, but the right to challenge these dates should be established from the beginning. However, executive override cannot cancel out the facts behind the constraints. It merely transfers responsibility for the consequences to the leader invoking that authority. And if both the valid concerns of the team and the override decision by the leader are documented, it provides clarity on risk ownership and ensures that if trade-offs are to be made, they're made with eyes wide open instead of blind optimism.

Implementation:

- Record assumptions underlying all constraints to the timeline.
- Establish escalation paths for constraint issues and who the ultimate decision maker is.
- Determine the true reason for constraints (capacity, knowledge, or speed) because each one requires a different solution.
- Face every challenge armed with options for resolution rather than presenting open-ended problems.

Example in Practice:

One utility company was pushed to launch a new customer relationship management (CRM) system prior to the fiscal year's end, a deadline imposed by executives without speaking to the implementation team after requirements-gathering. Faced with this deadline, the project manager did not simply acquiesce. She instead recommended deadlines in phases, the first being a three-month deadline to gather the requirements and perform software and vendor selection. After this, the goal would be set for implementation, separate from the goals of UAT, training, and go-live, which were to be established at the end of the subsequent phases.

With this level of detail, she walked into the meeting with a realistic breakdown of what was possible within the timeline, what would need to be deferred, and what further resources would be needed in order to complete the full scope. This fact-based approach led to a renegotiated plan with phased deliverables that ultimately succeeded, while the original all-or-nothing approach would have ended in failure.

One practical example is to bring options with you when facing timeline conversations: "If we eliminate these features, we can meet the deadline," or "With two additional developers, we can stick to the schedule," or "A three-week deadline extension would save us $X in added resource costs." Coming armed with options places the right decision in the appropriate hands.

8. The Right to Celebrate Small Victories

"We are permitted to celebrate all victories,
no matter how small."

It is all too easy, in the world of technology and project delivery, to have eyes fixed solely on the horizon, the looming final cutover, that decisive moment everyone anticipates with a mixture of hope and anxiety. Yet, by doing so, teams can overlook the vital progress being made along the way. I have seen a technical engineer excited to report, in a status meeting, that they have achieved migration of a sample data set.

Many times, tasks that seem small to the larger team can be met with muted enthusiasm and a look forward to a larger task that is yet to be completed. Allowing team members to bask in the glory of what they can report as success, rather than letting the shadow of future deadlines obscure present achievements, creates intentional celebration of these milestones. It is not only justified, it is essential. Acknowledging achievement keeps momentum and reinforces team structure.

Celebrating small wins signals that the project is on track, that systems are communicating, and that the foundational work for the larger task is solid. By pausing to recognize this success, teams see tangible results and receive acknowledgment for their efforts, making the daunting tasks ahead feel less insurmountable. This recognition encourages everyone to maintain momentum.

Implementation:

- Schedule milestone celebrations into the project plan.
- Make progress indicators visible to the entire organization to create transparency.
- Acknowledge individual and team efforts as a part of project status meetings.
- Share success stories with the broader organization using project newsletters or company-wide communication mechanisms.

Example in Practice:

One ERP vendor created a win wall on which they recorded achievements along the way throughout the project. At the end of every sprint, the team would meet to put up new wins on the wall. These were not necessarily technical accomplishments, but user adoption rates, process gains, and even difficult issues solved. This tradition maintained team interest through a three-year implementation and acted as a strong visual reminder of small successes during challenging times.

One practical application is to implement a good news segment at the beginning of every weekly project status meeting. Force each person to give one quick good thing that is going on in their project or personally in their life as a part of each meeting. This will not only begin every meeting getting the neurons of positivity flowing in the brain but will also open the door for everyone to celebrate together over small victories.

9. The Right to a Fast Pace

"We have the right to work at a fast pace where quality is done at speed and does not sacrifice it."

Finding the right pace for a project is an act of artistic finesse. Work too slowly, and teams get into a relaxed mindset that can cause projects to stagnate, and it becomes hard to speed up when the time is critical. Begin too intensely, and teams overlook important details and then incur the cost of rework and quality problems.

I have been a part of many projects that have had good people resign their position in the company throughout the course of the engagement because they burn out too quickly or because they feel like they are the only one sprinting while everyone else is content to walk.

In reality, the optimum speed is not comfortable. It's keeping the right amount of productive tension where teams stay engaged, alert, and able to tackle tough problem-solving. If things are too comfortable and predictable, you are not moving fast enough to drive change. You will know the appropriate pace when it is challenging, perhaps even a tad hectic at points, but not oppressive. Remember that a team can only move as quickly as its slowest critical member. Choose team composition thoughtfully, then create channels for members to share openly when they feel overwhelmed or underused.

Implementation:

- Deadlines should be aggressive and team members should be vocal about the pace.
- Track team workload and resolve overallocations on a daily or weekly basis.
- Prioritize quality metrics along with speed metrics. Ensuring quality is part of the success metrics; it's not enough to just check off the task from the list.

Example in Practice:

One practical application comes from Stephen Covey's well-known rocks, pebbles, sand, and water analogy and provides a powerful framework for task prioritization. In this model, the rocks represent the most important and impactful tasks, those critical priorities that, if neglected, undermine the entire project. Pebbles are tasks of moderate importance, while sand and water fill in with minor details and daily activities. Covey emphasizes that if you don't put the big rocks in first, they'll never fit later. Applying this principle means intentionally scheduling time and focus for the most vital work before less significant activities. By consciously distinguishing between what is essential and what is merely urgent or routine, teams and individuals can ensure that energy and resources are invested where they matter most, preventing the chaos of being perpetually occupied yet never truly progressing on core objectives.

A tech company that built business software add-ons had a key client who asked for an expedited schedule. The project manager offered this alternative: "Rather than dissipating our energy over several months, we'll put our entire crew on a two-week intensive burst with longer but concentrated hours, followed by a period of time off. We've found that this yields more than diffuse attention."

The team worked on a focused intensity model instead of a steady grind. Their teams put in incredible energy and focus only on rocks during core hours, completely there, collaborative, and engaged. But when they were finished, they were finished—no emails, no calls, no half-there family time still being mentally at work. Then, during the rest cycle, they focused only on sand and water activities, without the pressure of needing to make daily progress updates on the rocks.

Project measures showed that this "burn bright, then rest" cycle maintained quality and team enthusiasm without productivity loss from extended partial attention.

Not only did this result in improved client satisfaction through the client communication of the burn bright schedule, but it also kept the team intact and limited losses from past burnouts.

10. The Right to Change the Solution

"We have the right to revisit, reselect, or redesign the selected solution before or during implementation."

It is bordering on lunacy, that so many organizations commit to a software solution or vendor before meaningfully engaging in the very steps that determine the project's true needs: requirements-gathering, fit-gap analysis, and system design. This upside-down approach, where the cart is not just before the horse but has already left for the next town, seems almost normal to most senior leaders. Executives, seduced by glossy advertisements, vendor promises, or the allure of the latest tech ad on an airport wall, often convince themselves that this particular platform is the holy grail for every business woe. The project team is left to contort requirements, processes, and expectations into the neat box predefined by the vendor.

By selecting the software or platform before knowing what is truly required, organizations restrict themselves to solving the problems that a solution has already aligned itself to. It is as if a scientist, enamored with their hypothesis, forgoes testing or ignores the results, and publishes their theory anyway. In science, hypotheses are proposed, tested, and either validated or discarded based on evidence; progress is made only by embracing the possibility of being wrong about the solution and evolving accordingly. So too, in software selection, we must insist on evidence: rigorous requirements-gathering, thoughtful fit-gap analysis, and genuine system design. Only then can we say, with any confidence, that a particular software, platform, vendor, or add-on is the right solution.

Yet, in practice, the corporate world often does the opposite. The project team is told too late to change the solutions and to make it work with the current platform, regardless of misalignments uncovered during design or implementation. The result is a project that either limps along in mediocrity, rife with workarounds, bolt-ons, and frustrated users, or collapses altogether under the weight of unmet needs and unkept promises. The right approach, by contrast, is to grant project teams the explicit right, and the expectation, to revisit, reselect, or redesign the solution at any point before or during implementation. They must be empowered to say, "The requirements have evolved," or "This system does not fit," or "We need an add-on to bridge the gap."

This is not evidence of a poor past decision; it is wisdom born of learning and adaptation.

Flexibility must be baked into not just the technology, but into the mindset of the team and the expectations of leadership. Just as no credible scientist would cling to a disproven hypothesis, no responsible organization should lock itself into a software design before the evidence is gathered, interrogated, and tested.

Implementation:

- Establish scheduled solution reviews: Set regular checkpoints (e.g., after requirements-gathering, post-fit-gap analysis, or at major milestones) where the team formally assesses whether the various solutions still align with needs.
- Leverage prototyping: Build early prototypes throughout the design phase. If significant issues or gaps emerge, use this evidence to justify reevaluating the chosen platform before full-scale implementation.
- Start with a platform: Rather than selecting a software solution select a platform based on its ability to handle any designed solution. Tools that allow you to fully customize solutions without writing complex code can be a good tool for starting to iterate and prototype a solution.
- Contract in phases: Work with a vendor that will contract for only a phase of the project rather than contracting for the licensing and implementation at the design phase.
- Document and track unmet requirements: Maintain a living log of requirements that are only partially met or not met at all (fit-gap). Use this log as evidence during solution reviews, supporting objective decisions to revisit or redesign the system.

Example in Practice:

Consider the case of a major energy production company navigating the complex world of ERP selection. With only fifty employees, the leadership opted for the vendor's small-business tier solution, an apparently sensible decision based on headcount. However, a deeper dive into their requirements revealed the need for the robust features offered only in the vendor's enterprise-tier system. Because the software choice was locked in early through a price-driven RFP process, the project team was denied the opportunity to revisit this decision. As the realities of implementation unfolded, critical gaps became glaring. The chosen ERP lacked native support for several key HR processes that the company relied upon. Instead of reconsidering the entire platform, the team resorted to patching the system with an array of third-party add-ons, introducing complexity, cost, and risk that could have been avoided by aligning the software choice after actual requirements-gathering.

One practical application of this is to conduct a system selection process that goes beyond the typical high-level requirements and evaluate platforms that meet these requirements. Next, engage in a project to create the solution based on the detailed solution design and true fit-gap analysis process before making a decision and signing contracts for software licensing.

Executing the Project Bill of Rights

When embarking on a complex technology project, it's essential to acknowledge that the landscape will inevitably shift as requirements are clarified, priorities evolve, and new information comes to light. For this reason, many of the principles that underpin a successful Project Bill of Rights depend on the ability to adapt to change, not only in solution or timeline, but also in cost.

Rather than holding vendors or implementors to a rigid, predetermined price for each phase, consider the value of embracing a more flexible approach. Cost ranges, thoughtfully structured around phases, recognize the unknowns that are part and parcel of the project evolution. This flexibility invites more honesty from partners, who are then free to estimate based on current knowledge, without the pressure to disguise uncertainty behind prices that rarely last during the pressure test of the engagement.

When we ask for figures too early, we may unwittingly encourage the creation of numbers that are, at best, guesses. Worst case, the vendor closes off potential solutions because they have been designed to cost rather than actual requirements. By opening the door to transparent discussion of possible outcomes—what might happen if requirements evolve, technical complexities arise, or the business context changes—we build a sense of trust and realism across all parties.

In this way, a project's cost structure becomes as much a living part of the process as the solution, able to flex in response to lessons learned. Accepting this level of flexibility is not a sign of weakness or lack of control; rather, it demonstrates a mature understanding of how true value is created.

Drafting a Project Bill of Rights document is only the beginning. To bring these tenets to life, consider the following implementation strategies:

1. Make it official
    Include the Project Bill of Rights in your project charter or contracts, and get all of the stakeholders to sign off on it. This transforms it from a nice idea into a binding contract that sets up project governance. Print a copy to keep on the table of any project conference rooms used during the project.

2. Reference it regularly
    Refer to specific rights at the appropriate times during discussions. If someone is hesitant to raise a point, remind them of the right to constructive disagreement. If scope changes are proposed without changes to the timeline, invoke the right to make such changes.

3. Lead by example
    Leaders need to exemplify these rights in their own actions. This could include publicly admitting when you're wrong, recognizing team members who speak up on contentious issues, or standing up against unjustified requests.

4. Create supporting processes

Each right requires operational processes to render it concrete. For instance, the right to scope changes requires a formal change management process with impact analysis.

6. Measure and improve
   Periodically gauge how well your team is doing in living up to these rights. Anonymous surveys will tell you if team members really do experience psychological safety, if they think they can confront unrealistic constraints, and if mistakes are being viewed as learning opportunities.

From Dysfunction to Success

Janet Chen, the new CIO of a regional bank, spoke to the executive team. On the screen behind her were the dismal figures of the bank's CRM project, the third failed technology project in two years.

The CEO leaned forward with barely contained outrage, "So you're telling me that we've spent $4.2 million and we've got nothing? And now you want more money and time?"

Janet didn't bat an eyelid. She had been brought on board to do precisely this, to revamp a firm with a history of over-promising and under-delivering.

"What I'm telling you," she replied, "is that we've been solving the wrong problem. We've been treating this as a technology issue when it's actually about creating an environment where people can speak honestly about challenges, timelines, and risks without fear of punishment. No amount of new code or different systems will fix a culture where people are afraid to tell the truth."

The room fell silent. Janet moved to her next slide, which showed an anonymous survey of the project team. Eighty-seven percent indicated that they were aware of major issues that they felt were not being addressed. Ninety-two percent indicated that the timeline set by the executive team was unrealistic from the start.

"Your people have lied to you," she continued, "but not because they wanted to. Because they had no choice."

She depicted a workplace in which messengers were killed, speaking up was career suicide, and impossible deadlines were imposed with the clear message: make it happen or look for another job.

"Your former CIO fostered an environment of fear," Janet went on, and gave examples. "The integration architect was pulled from the project for raising compatibility questions, the database group was told 'just get the data in' when they requested additional time for data migration, and the security lead was labeled 'not a team player' when she voiced compliance shortcomings."

The CFO cut in, defensively. "We require accountability around here, not excuses."

"Accountability without the ability to voice concerns is not accountability—it's terror," Janet replied. "And terror produces silence, not success."

She put a paper on the table. "This is what I suggest. Before we write one more line of code or spend one more dollar, we lay a new foundation. I call it our Project Bill of Rights."

Janet described each right in detail throughout the next hour. When she finished, the room was quiet. Then the CEO spoke.

"This feels soft. We're a bank, not a therapy group."

Janet smiled. "I anticipated that reaction." She clicked to her final slide, revealing implementations of similar strategies at organizations like NASA, Toyota, and other Fortune 500 companies, along with their metrics of success.

"This isn't soft; it's data-driven leadership. NASA discovered that high-psychological-safety teams detected 58% more potential mission-critical failures pre-launch. Toyota's 'stop the line' culture, basically the right to speak up without fear, cut defects by 80% while improving productivity."

The Project Bill of Rights is not about moving slowly or avoiding difficult conversations. On the contrary, it creates the conditions under which people speak more, engage more, and actually work much harder because they have confidence that everybody is moving forward together with intent. When they feel that they are valued to contribute and speaking up will not be punished, they invest their entire energy into outcomes rather than spending it in self-defense. The result is more momentum and speed, not less, as everyone on the team is fully committed rather than half-hearted.

The VP of Retail Banking, who so far had not uttered a word, leaned forward. "My business lost three key customers due to the last CRM debacle. At this point, I'll do whatever it takes. You have my backing."

One by one, the executives tentatively agreed. The Project Bill of Rights was officially adopted the following day.

The change did not occur overnight or gracefully. The initial safe space conversation was greeted with doubt and uncomfortable silence. But Janet dissolved the tension by sharing her own mistakes on a past project, leading by example to make it clear that honesty was genuinely encouraged.

Gradually, information trickled in. One of the senior developers came forward and shared that the data migration methodology had underlying flaws with the way it migrated record keys. A business analyst owned up to leaving out what the department leads called critical requirements because a key executive stakeholder had downplayed them. The QA lead owned up to reporting green status on some major tests despite being blocked from continuing with tests due to code not being complete.

Instead of blame, each discovery was greeted with solutions. Teams were redirected to address these freshly uncovered problems. When scope was found to be 40% greater than original estimates, Janet didn't request the team to labor more diligently; she arranged a prioritization session with business stakeholders to tackle the highest-value components first and deprioritize other less valuable items.

The turning point was three months in. During a steering committee session, a young developer cut into the CTO's status report to point out a glaring security flaw in the proposed architecture. The room grew quiet, bracing for the usual backlash. Instead, the CTO thanked her and ordered a team to analyze the problem on the spot.

Word spread quickly: this wasn't just talk, it was real. Issues that had festered for months suddenly came to light. The project plan was completely restructured around reality rather than wishful thinking. Six months later, the initial phase of CRM went live, not the complete system that had initially been promised, but a foundation that provided the most important capabilities without a hitch. Customer satisfaction ratings for functionality rose 28% during the initial month after launch.

Over the celebratory lunch, the CEO pulled Janet aside. "I didn't think this would work," he admitted. "I thought we needed a harder edge and more pressure."

Janet motioned to the team, now engaged in a spirited conversation about the next phase of the project. "Pressure doesn't always create diamonds in software projects; it creates imperfect carbon. These people always had the potential to do great work. What they needed was the safety to speak the truth and the permission to do quality work without the fear of changing the cost and the timeline."

The methodology extended beyond the CRM project. The bank applied the Project Bill of Rights to every technology project within a year and then to other departments. Project success rates went from 20% to 75%. Employee retention within projects improved significantly. The bank's Net Promoter Score™, which had been one of the lowest in the region, moved upward steadily as stable, quality systems took the place of faulty, rushed implementations.

The primary lesson was infused in the culture of the bank: psychological safety is not an emotional notion; it's required for project success. When individuals are able to speak truth without fear, issues are solved prior to turning into disasters. When groups are shielded from unrealistic expectations around timelines or budgets, they produce realistic outcomes. And when quality comes before speed, both ultimately excel.

And, as Janet liked to remind incoming project managers: "The technology didn't change. The team didn't change. What changed was building an environment where people could do their best work, where honesty was more important than comfort, where issues could be solved before they turned into disasters, and where everybody was working for project success instead of arbitrary dates and unrealistic budgets."

Moving Forward

As you get ready to organize stakeholders and bring your implementation team together, the Project Bill of Rights paves the way for success. It not only establishes what you'll build, but how you'll work together to build it.

Reflection Questions

1. Which of the rights delineated in the Project Bill of Rights would have the most impact on your organization's project if implemented today?
2. Which particular behavior in your organization presently violates these rights?
3. How would you present the idea of a Project Bill of Rights in order to gain stakeholder buy-in?

4. Which right would be most difficult to apply in your company, and why?
5. How would you modify or expand on this Bill of Rights to respond to particular issues in your project environment?
6. Which personal behaviors would you need to change to effectively model these rights?

CHAPTER 5

Aligning Stakeholders for Long-Term Success

"Alone we can do so little; together we can do so much."
—Helen Keller

I n the conference room, tension filled the air as the status meeting on the project broke down into another exercise in blame.

The Marketing Director chimed in, her voice tinged with frustration: "The marketing group requires this data and functionality, and we've waited three months."

"We already told you that's not feasible with our current architecture," the IT Director replied flatly.

"Had you spoken with us prior to choosing your own platform, we would not have these limitations that require significant customizations."

The CEO, who had been quietly listening, finally spoke. "This is precisely why we're six months behind schedule and four times over budget. We've got departments working against one another rather than toward mutual objectives."

This scene plays out in organizations worldwide, revealing a fundamental truth about digital transformation: technical challenges are solvable with the right expertise and resources, but misaligned stakeholders will derail even the most technically sound implementation.

The Importance of the Leadership Team

While senior leadership forms the foundation of the project, the building blocks of the broader implementation significantly impact success. Beyond filling specific technical roles, organizations must consider three critical dimensions of team composition to ensure skills, personalities, and knowledge are achievable.

The most effective implementation teams are assembled based on needed skills rather than organizational titles or departments. By doing this, you must:

- Focus on team member capabilities required for success
- Eliminate departmental silos and internal corporate politics
- Ensure all critical skill areas are represented across the team
- Create teams that solve problems rather than defend turf

Achieving Stakeholder Alignment

Stakeholder alignment is the solution to complexity problems. When everyone is aligned, working toward common, measurable objectives, organizations avoid scope creep, role confusion, and communication breakdowns. Clear purpose enables teams to navigate the inevitable complexities of large-scale digital transformation.

Having stakeholder alignment is also critical to resolving human problems. By engaging stakeholders in structured participation, organizations intentionally bring problems to the surface and solve them, work to establish shared ownership, and kill power struggles before they happen. Clearly defining roles, such as the project visionary and project integrator, establishes successful leadership-team dynamics. Capability-based teams work beyond territorial disputes by focusing on what role they play instead of job titles.

In a way, building a team based on skills that are focused on a common set of aligned goals creates the foundation of understanding, purpose, and accountability necessary for true digital transformation success.

The Tale of Two Approaches

It was the best of implementations; it was the worst of implementations...

At one manufacturing company, the IT Director was an arrogant tyrant. "The IT department knows what is best when it comes to selecting software," he'd snap, dismissing concerns from the shop floor to the executive suite. He had chosen a sleek new ERP system that was loaded with enterprise-grade bells and whistles and a vendor who guaranteed a smooth rollout. The consultant he had recommended promised extravagant efficiencies.

Eighteen months later, the shop floor resembled a scene from a sci-fi film. Managers stood around huddled over spreadsheets, manually recording production numbers because the new system couldn't handle their job travelers. The salespeople were inundated with angry customer emails, having no means of determining if parts were in stock or on back-order. Finance burned the midnight oil reconciling invoices between the ERP and their Excel-based trackers. The digital paradise that had been promised by the IT director was instead a landscape of workarounds and team members with barely contained rage.

Meanwhile, at another organization, the IT Director practiced true leadership. She sat alongside machinists to learn about their work order process, shadowed salespersons to learn about customers' pain points, and analyzed the details of reporting needs from accounting. Under her leadership, every department selected software in a silo according to the department's needs.

But they suffered a similar crisis the following year. The various disconnected systems and highly customized integrations created hazardous data issues. Moving orders for sale to production was a manual exercise in data exporting, transformation, and importing separate systems. Inventory tallies never balanced in the warehouse.

All cross-functional processes were hazardous tightrope acts across system boundaries that required technical team members to be available all hours of the day and night, at the beck and call of all departments who may need integrations 'run' or stuck transactions 'retried.' The bespoke solutions had left the organization in worse shape and required more people to operate than before they made the changes.

Imagine all departments speaking a different language: sales monitored opportunities in one framework; manufacturing had an entirely different data schema called quotes; and finance recorded draft transactions in yet another framework. The data could not easily convert from one to the other, since they named and described critical information differently. What sales called a "customer order number" would be a "work order number" to production and an "accounts receivable ID" to finance. Without expensive custom translation software (integrations), employees became human translators, typing information back and forth between systems and manually moving failed transactions. This virtual language barrier presented endless opportunities for errors, copies, and lost information, turning what otherwise would have been streamlined digital processes into error-prone, time-consuming tasks.

In these contrasting tales, we see the perils of both extremes: The dictatorial approach that imposes ill-fitting systems, and the overly deferential method that creates a jumble of disjointed solutions. The common thread? A failure to align stakeholders and leadership effectively.

True digital transformation success is all about achieving a delicate balance, offering strategic direction and embracing varied requirements. It involves energizing everyone toward common goals, balancing visionary leadership with methodical execution, defining teams in terms of required capabilities instead of fixed positions on an org chart, and offering formal channels that require stakeholder input.

In the next two chapters, we'll explore the main methods of ensuring this alignment occurs. By establishing clear goals, separating visionary and integrator roles, forming capability-based teams, selecting the right consulting partner, and ensuring regular stakeholder engagement, organizations can establish the mutual commitment that enables successful change to take place. Let's look at each method in turn.

Developing Clear Objectives and Project Charters

The cornerstone of stakeholder alignment is clarity of purpose. Without clear, shared objectives, stakeholders inevitably pursue conflicting priorities. What may seem like clear objectives to senior leaders who are inexperienced with large-scale technology projects will prove to be half-baked and idealistic words that cannot translate into actionable plans as the implementation progresses.

Consider this example from a global retail chain implementing a new inventory management system: "We want real-time inventory visibility across all locations."

This kind of dangerous ambiguity appears constantly in RFPs and requirement documents across many organizations. You'll see statements like "The system shall have real-time visibility to inventory" without that crucial qualifier capturing the actual business value or outcome. The warehouse worker may interpret this requirement to mean anyone in the company can launch the ERP system and see inventory levels for any item across the various warehouse locations. The account managers believe this will give them the ability to see all availability on their mobile phones while they are in a retail store. The marketing team may believe this to mean that the website will allow customers to search for inventory across all the various stores to easily determine who has a product in stock, place the order, and support store-to-store transfers of inventory where required.

Consider this modification to the language: "The system shall have real-time visibility to inventory so that customers can see on the website whether a potential product is in stock, at which location, or if it is back-ordered." Without connecting technical capability to customer experience, all stakeholders interpret requirements through departmental lenses, leading to disappointment when the implemented system meets everyone's needs poorly rather than meeting anyone's needs well. The best digital transformations always connect technical needs to tangible, measurable customer results.

Without alignment, the implementation team for this requirement found themselves trying to satisfy imaginary requirements, leading to an overcomplicated system that satisfied no one.

A better approach identifies specific outcomes with measurable metrics:

- Reduce out-of-stock situations by 15% within six months through predictive inventory.
- Decrease inventory carrying costs by 8% within one year through store-to-store transfers.
- Enable store staff to locate any item in under 30 seconds through location tracking.
- Provide 98% accuracy in available-to-promise calculations on the website.

This level of specificity leaves little room for misinterpretation and provides clear success criteria for evaluating progress.

To develop clear, aligned objectives, start with business outcomes, not features or requirements:

- Focus on measurable business results rather than technical specifications.
  - *Example*: Instead of stating, "Implement a cloud-based CRM," say, "Increase customer retention by 15% through improved relationship tracking and proactive engagement."
- Ensure objectives address all stakeholder needs.
  - *Example:* For an order management system, include outcomes for customers (faster delivery confirmation), sales

(accurate inventory availability reporting), warehouse (optimized picking routes), and finance (automated revenue recognition).

- Identify what success looks like for each department and stakeholder group.
  - *Example:* For HR, success might be, "Reduce onboarding paperwork time by 75%" for new employees. "Complete all required documentation in a single 30-minute session before first day of work."
- Quantify success wherever possible.
  - *Example:* "Improve efficiency" is too vague. "Reduce processing time from 45 minutes to 15 minutes" is clear.
- Establish priority and sequence. Not all objectives can be achieved simultaneously; clarify which come first.
  - *Example:* Phase 1: Implement core inventory management to raise visibility of inventory stockouts to every person in the company; Phase 2: Add automated replenishment to optimize cash flow by 10%; Phase 3: Integrate customer-facing inventory visibility to the website for all customers.
- Clearly document and communicate objectives to all leaders and departments to ensure mutual understanding. This transparency helps departments see the reasoning behind changes or sacrifices they may need to make, fostering cooperation and buy-in.
  - *Example:* Create a one-page Project Outcomes Letter from the CEO visible to all departments showing how each business unit benefits, what sacrifices are needed, and the company-wide and client-side impact of the transformation.
- Create artifacts that clearly articulate objectives and refer to them consistently.
  - *Example:* Develop a North Star statement and document, linking each system capability to specific business outcomes with clear measurable goals or KPIs, and review progress against these metrics in every status meeting.

These well-articulated project objectives become the foundation of your project charter, a document that serves as the single source of truth for project purpose and scope. An effective charter will include:

- Problem statement: What specific business challenges are we addressing?
- Business case: Why is this investment worthwhile?
- Success criteria: How will we measure achievement?
- Scope boundaries: What is explicitly included and explicitly excluded?
- Key stakeholders: Who is involved and what are their roles?
- Timeline and milestones: What is our roadmap for delivery of the first phase?

A charter developed collaboratively and endorsed by all stakeholders becomes a powerful tool for maintaining alignment throughout the project lifecycle. If this key foundation is not built, even the technically flawless implementation will be deemed a failure. With that in mind, the following are worth keeping in mind when developing the charter and setting engagement expectations:

1. Clear articulation of project objectives and deliverables:
   a. Expectation management begins with precise articulation of what the project will and will not deliver, and the boundaries.
      i. What is explicitly included in scope?
      ii. What is explicitly excluded from the scope?
      iii. What might be considered for future phases?

   b. Phased delivery expectations
      i. Clear milestones with defined outcomes
      ii. Understanding of incremental value delivery
      iii. Realistic timelines to achieve benefits and what to expect from each phase

2. Clear communication that aligns: Even with clearly articulated objectives, alignment fails when different organizational levels

operate with conflicting priorities. A successful transformation requires vertical leadership, management, and operational goals alignment from strategic leadership through tactical management to operational execution.

   a. Creating multi-level alignment requires:
      i. Cascading departmental goals
         1. Link every feature and process decision to strategic objectives.
         2. Translate strategic goals into department-specific success metrics.
         3. Connect department goals to daily operational improvements.
      ii. Cross-departmental priority alignment
         1. Facilitate open discussion about department-specific priorities.
         2. Identify and resolve inherent conflicts between departmental needs and how they impact other departments.
         3. Establish clear trade-off decisions and request compromises.
      iii. Continuous Communication
         1. Regularly revisit expectations as implementation progresses.
         2. Address emerging gaps between expectations and reality.
         3. Adjust delivery plans based on evolving organizational needs.

Achieving Buy-In to the Charter

A developed charter without buy-in from the entire organization creates a façade of agreement that will crumble under the pressure of an implementation. True buy-in transforms stakeholders from reluctant participants into active champions, driving the change even when the implementation cycles get challenging.

Look at this example of what occurred at a financial services company rolling out a new customer service platform.

The Project Manager sternly advised the consulting team that the CEO "absolutely hated AI" and had "vowed to never permit it within the organization." Accepting this at face value, the consultants went out of their way to omit any AI-related capability in their design, not even discussing these alternatives for fear of inciting executive ire.

Half a year into implementation, during a project dinner, the CEO casually talked about utilizing AI tools to draft personal communications and to assist in reading board reports. Surprised, the lead consultant diplomatically inquired about his thoughts on AI. "It's changing how I work," the CEO responded enthusiastically. "I've been experimenting with it for months. We should be looking at leveraging this across the business."

The revelation came as a surprise. The CEO had formed his own opinions about AI based on personal experimentation, but the change of mind had never been communicated to the company. The project team had designed a totally AI-free solution based on outdated assumptions, missing the opportunity to incorporate capabilities that would have revolutionized the value of the platform.

This lack of alignment set the project back three months and almost $400,000 in redesign. The lesson was obvious: alignment is not a single checkbox but a continuous conversation. Stakeholder views change, attitudes shift, and assumptions need to be continuously verified instead of being assumed as facts. The consulting team was too willing to take the Project Manager's interpretation of where the CEO was, instead of seeking opportunities to discover these opinions firsthand and continuously check in and validate.

Here are some strategies to involve stakeholders at the various project phases:

- Project initiation and planning
  - Conduct organization hierarchy analysis to identify all affected stakeholders.

- o Establish inclusion frameworks that respect time constraints while ensuring representation. You may only have 15 minutes to brief some team members, while others may be supportive of a deeper analysis within the project.
  - o Create multiple input channels for different stakeholder types.
- Design and configuration
  - o Implement iterative design review cycles that quickly highlight successes and challenges that are occurring within the engagement.
  - o Use interactive prototyping to make concepts tangible.
  - o Facilitate structured sessions with clear parameters for feedback, and test case results that are communicated prior to the meeting.
- Testing and validation
  - o Involve actual end users in acceptance testing; go beyond the project team.
  - o Create realistic test scenarios based on day-to-day operations.
  - o Provide structured feedback mechanisms to log all usability concerns.
- Implementation and go-live
  - o Train and empower departmental champions who can carry the message and goals across their departments.
  - o Create transparent escalation paths for critical issues that eliminate any potential behind-the-back reporting of issues.
  - o Establish ongoing feedback loops for continuous improvement.

Moving Forward

Alignment to objectives and stakeholder communication is the basis for effective digital transformation. Without it, even the most technically proficient of implementations won't be able to provide expected value. The principles we've outlined here—a clear definition of objectives, an appropriate definition of roles, formal stakeholder engagement, management of expectations, and inclusive buy-in processes—establish the human infrastructure necessary for transformation success.

Reflection Questions

1. How clearly articulated are your project objectives? Could every stakeholder give consistent answers about what success looks like?

2. Is your implementation team composed based on organizational titles or needed skills? What gaps might exist in your current team structure?

3. What formal and informal channels exist for stakeholders at all levels to provide meaningful input?

4. What mechanisms exist to ensure honest feedback about implementation concerns?

5. How do you balance the need for structured stakeholder involvement with respect for participants' time and primary responsibilities? Do all stakeholders participate in unique channels of communication based on their role and commitment to the project?

6. What evidence would indicate that stakeholder alignment is improving or deteriorating in your current implementation?

CHAPTER 6

Choosing Project Leadership Wisely

> *"If you want to build a ship, don't drum up the men to gather wood, divide the work, and give orders. Instead, teach them to yearn for the vast and endless sea."* —Antoine de Saint-Exupéry

As we have discussed in prior chapters, the posture of the leadership team and the individual personnel is the cornerstone of any effective digital transformation project. Companies likely will get bogged down in technology selection, requirements definition, project planning, and budgeting if they underestimate how much the project leadership team's decisions will be the key to day-to-day success within the project. Time and time again, we've witnessed technically accurate deployments fail as leadership dynamics develop decision paralysis, competing agendas, or simply turn over their staff in a way that becomes detrimental to the engagement.

Digital transformation initiatives require more than technical know-how. They need the correct leadership framework to oversee the intricate dance among departments, individuals, processes, and technology. In this chapter, we will discuss the most critical leadership roles that need to be filled in any successful transformation, how to find the correct individuals for those roles, and how to organize teams for optimal performance.

Key Roles: Visionary vs. Integrator

In their book *Rocket Fuel*, authors Gino Wickman and Mark C. Winters describe a powerful leadership dynamic that can be applied perfectly to digital transformation projects (Wickman, 2016). While they masterfully apply this model to organizations as a whole, the same principles transfer remarkably well to the world of digital transformation projects. In Rocket Fuel, they define two unique leadership personalities that, when combined appropriately, form what they refer to as rocket fuel for organizational success: the visionary and the integrator.

Knowing the dynamic between these two roles, what makes each one special, and the limitations of each is essential because many project failures result from one of these three scenarios:

- Lack of a clear distinction between these roles
- Having the wrong individuals in these roles
- Fundamental misalignment between these leaders

To gain a better understanding, these two stories contrast the visionary and integrator leadership styles in different scenarios. Both stories highlight how these different, but complementary, leadership approaches work in practice, and particularly how they need to work together for successful project implementation.

The Visionary	The Integrator
The CMO at a chain of retailers was chartered to spearhead their omnichannel customer experience overhaul. A natural visionary, he painted a vivid picture of customers flowing effortlessly between online and in-store interactions, with suggestions tailored to their interests, live inventory	When the new quality control implementation began to unravel, the company knew that it required an integrator. It assigned the Director of Quality, a methodical, detail-driven executive respected for her ability to get things done, to the

availability, and a seamless checkout experience.

His presentations to the board were convincing. He secured a large budget and created enthusiasm for the project. In design sessions, the ideas flowed out of him—mobile app functionality, loyalty program integration, virtual reality dressing rooms, social media commerce functionality.

As a textbook visionary type, he constantly came up with new ideas, but none of the critical foundational priorities were getting done. Once the initial presentations were over, the vision cast, and the work shifted to detailed requirements-gathering, he lost interest. His mind and enthusiasm had already moved on to the next exciting project, leaving the team without direction on execution priorities.

IT staff and store managers were excited at first but then became frustrated. Requirements continuously shifted. Scope inched along endlessly. The CMO became impatient with technical position of full-time integrator for the project.

Her initial steps were thoughtful. She did a thorough scope review, listing all the additions that had crept in over the past few months. She separated them into must-haves for initial launch versus future enhancements for later stages. She defined concise decision criteria for any scope change proposals.

Most importantly, she established a regular cadence of meetings: daily standups to address near-term problems, weekly status meetings to maintain alignment, and monthly steering committee meetings where the visionary VP was able to offer input and guidance on strategic choices.

The project was back on schedule in two months' time. However, when the VP began to look into the original goals and objectives of the project, it became clear that the new system was going to be a lift-and-shift of the existing system and offer no new functionality. Innovative vision and any kind of new strategic direction were

limitations and implementation timeframes. Nine months in, they had spent 80% of the budget and had nothing deployable. Teams were demoralized, and executive leadership was wondering if the project was even viable.	missing. While the Director ensured that planning translated into actual progress on the ground, her ability to offer a vision that would prove transformational was lost in the day-to-day shuffle of tasks.

With a basic understanding of each role, let's examine each role in greater detail.

The Visionary: Driving Creativity and Vision

The visionary role within the project serves an invaluable function that cannot be minimized. You must realize that the visionary is not a role assigned merely within a project setting, but rather a personality type that naturally exists within a business. They are usually assigned leadership positions precisely due to their charismatic personality and their talent for creating compelling visions that motivate others.

The visionary plays a crucial role. This leader:

- Articulates the why behind the transformation
- Paints a compelling picture of the future of the organization or environment
- Connects the individual project to broader strategic objectives of the company
- Drives innovation and creative problem-solving
- Inspires and motivates teams through challenges
- Maintains energy and enthusiasm throughout the project

However, visionaries also typically:

- Generate a constant flow of new ideas that can expand or derail the objectives

- Lose interest in implementation details and the nuance of complications
- May change project direction frequently based on latest insights
- Grow impatient with necessary process steps, feeling like things are moving too slowly
- Can underestimate task complexity and implementation challenges

The visionary is the lifeblood of the project. They can come up with new ideas in a matter of moments, but they can also pull the project off track, make it go over budget, and change the scope daily if not held in check. Many times, because they are in senior leadership roles, founders, or in senior management, most organizations believe the visionary should lead the project. They are charismatic, persuasive, and have the energy that gets stakeholders excited. Their presentations receive budget approval and their passion is contagious. Even if not placed in direct responsibility, they have a way of wrangling control of a project by simply commanding the room and bending everyone to their will. Without structure and balance, the same energy becomes detrimental to project delivery. This outcome must be resisted, which means you can't have a visionary leading the project.

The visionary's strongest attributes—creativity, strategic vision, the ability to inspire trust and excitement, and the capacity to envision possibilities—become weaknesses when not balanced by the counterweight of a successful integrator.

The Integrator: Ensuring Execution and Focus on Deliverables

The integrator complements the visionary by translating the vision into reality. In any project, the role of the integrator is just as important as that of the visionary. The integrator is the counterweight to the visionary. This leader:

- Ensures systematic implementation of the vision

- Maintains core focus on deliverables and timelines
- Manages project scope to prevent creep or drift from objectives
- Coordinates across departments and functions because of a deep understanding of all business units or departments involved
- Translates between technical design and business processes
- Creates the accountability layer for execution
- Establishes and maintains meetings and processes that drive steady progress

The integrator personality excels at:

- Ensuring follow-through on commitments by even the most senior stakeholders
- Maintaining project discipline and meeting cadences
- Converting abstract goals into concrete actions and plans
- Marking things "done" to an obsessive level
- Gracefully saying "no" or "not yet" to scope expansion
- Keeping teams aligned with project objectives by lifting up the vision of the visionary

But integrators may struggle with:

- Motivating teams when morale is low or project objectives are drifting
- Adjusting to required changes as organizational conditions shift
- Seeing the big picture outside of immediate tasks and deliverables
- Balancing their focus on process versus outcome, tending to favor the former
- Appearing to be resistant to innovative solutions that contradict prior plans
- Communicating the why behind choices made with the visionary

That integrator personality gets amped up about taking the vision from the visionary and implementing it, getting it down on paper so that the visionary can see progress. They take responsibility for scope, for budget, and maintaining oversight of key project objectives. These are critical factors in seeing the project objectives become a reality.

Balancing Creativity and Execution

The most successful transformations occur when visionaries and integrators work in concert, each respecting the value the other brings.

In projects where this partnership works well:

- Visionaries are free to come up with new ideas without getting bogged down in implementing end-to-end solutions.
- Integrators have clear direction and authority to execute the vision.
- The overall team benefits from both the inspiration and structure of having someone who has permission to say "no" to the new ideas.
- Scope remains aligned with outset strategic objectives.
- Execution proceeds methodically while accommodating necessary adjustments.

This collaboration sets up a dynamic synergy whereby vision drives change and implementation guarantees results. Neither role attempts to perform both functions, a pitfall that results in either diffused vision or a non-innovative implementation. When the roles are spoken aloud, there's a shared sigh of relief that each can concentrate on doing their role. Visionaries do not feel weighed down by implementation details, and integrators can concentrate their organizational skills without the need to come up with inspiring visions. There is a freeing feeling that comes when this dynamic is transparent within a project.

However, as power struggles and egos many times do, if this cooperation declines, predictable dynamics ensue. Take these real-world examples:

- Visionary without integrator: Exciting kickoff, expanding scope, initial progress, eventual stalling as details and follow-through are neglected

- *Example:* A CEO of a manufacturing firm launched an initiative for a virtual reality product visualization tool with much fanfare and support from the organization. Six months later, they had prototypes of touchscreen interfaces for the factory floor, augmented reality-based training concepts, and device sensor prototypes, but not one production-ready piece. Vision continued to expand from the initial idea, but execution lagged, and engineers were drawn in different directions each week, depending on the CEO's newest idea.
- Integrator without visionary: Efficient execution of a plan that may not inspire or connect to the strategic needs; tasks completed but opportunities for innovation missed
  - *Example:* A COO at a financial services company spearheaded an initiative for workflow automation that painstakingly digitized current client onboarding processes, including their inefficiencies and flaws. The team met all the deadlines and was below budget, but the resulting system merely replicated their paper processes electronically, failing to take the opportunity to redesign customer experiences or leverage new technologies implemented by the competition. A year later, they were even further behind their leading competitors.
- Misaligned partnership: Friction between project leadership, conflicting direction to teams, political struggles, and eventual project derailment
  - *Example:* At a health care provider, the CIO (integrator) and Chief Medical Officer (visionary) both laid claim to a patient engagement portal. The CMO made promises to physicians about new functionality the CIO hadn't signed off on, and the CIO spoke of technical limitations the CMO hadn't been consulted about and did not fully understand. Teams were getting contradictory directions on a weekly basis, progress meetings devolved into territorial bickering, and after $2.3

million of expenditure, the project wound up canceled without any kind of deployment.

Implementing the Visionary-Integrator Dynamic in Your Project

The challenge is to find and develop leaders who are able to span the vision-action gap. Let us deal with practical ways of discovering and implementing the visionary-integrator dynamic in your project.

Among the most frequent questions that organizations have is how they can spot potential visionaries and integrators within their current organization and project. Not all transformations can afford to recruit new leaders for these positions.

Recognizing Natural Visionaries

Natural visionaries are often:

- Creative problem-solvers who generate ideas for solutions
- Strategic thinkers who see connections to larger business goals others do not see
- Inspirational communicators who can articulate compellingly
- Risk-tolerant personalities who are comfortable in ambiguity
- More interested in what's next than in current operations
- Not particularly interested in detail or impatient with documented processes
- Energized by new possibilities that can revolutionize their industry or organization

For those curious about their own tendencies, *Rocket Fuel* includes an online assessment that helps determine whether you naturally lean toward being a visionary or an integrator. This can be a valuable starting point for understanding your natural strengths and how you might partner with complementary team members.

Consider assessing potential visionaries with questions like:

- "How do you envision this change will transform our organization three years from now?"
- "What other opportunities outside of our core business should we be seeking?"
- "What excites you most about this project?"
- "How do you think this project aligns with our overall company strategy?"

Their answers can reveal whether they naturally think in visionary terms.

Recognizing Natural Integrators

Natural integrators typically are:

- Strongly focused on execution and relentless follow-through on even simple tasks
- Likely to apologize if they are even one minute late to a meeting
- Check-list driven personalities—They may add something to their list that is already done just so they can receive the satisfaction of marking it complete
- Process-oriented with such attention to detail that they may struggle at first to visualize the abstract
- Able to organize complex initiatives quickly
- Skilled at translating abstract concepts into action plans
- Comfortable and convincing when saying "no" to even the most senior of leaders
- Talented at coordinating across departments
- Innately pulled to complete open initiatives before starting anything new

Assess potential integrators with questions like:

- "How would you approach breaking this vision down into executable steps?"

- "What systems would you establish to ensure consistent progress is being made?"
- "How have you managed scope and definition on previous projects?"
- "How can you balance competing priorities from different stakeholders?"

Identifying these roles often happens naturally. Most of the time, after only a couple of project meetings, we can spot who our integrator is and who our visionary is because the personality differences are so drastic.

Once identified, leaders often benefit from coaching to maximize effectiveness in their natural role:

Coaching visionaries:

- Channel creativity into structured innovation meetings where they can use their imaginations.
- Develop discipline around communication timing and to whom they communicate vision.
- Establish clear expectations for their involvement and what the boundaries are for their ability to innovate within the organization.
- Help them recognize the value of process and structure as a mechanism to see the vision through to reality.
- Create appropriate forums and solicit their input periodically. Even if nothing changes in the scope, this outlet will be important.

Coaching integrators:

- Encourage strategic thinking and broader vision alongside tactical execution.
- Develop relationship, comfort, and trust with the visionary..
- Build their skills for managing visionary relationships and how to handle conflict.

- Help them communicate the why from the visionary alongside the how that comes naturally to their minds.
- Don't go around them. Support them in saying "no" effectively when needed.

You may not be able to do formal role designation in many projects because of hierarchies or pre-existing roles. Where direct assignment is not feasible, subtle influencing and beginning to use these terms within the project team is critical. Behind-the-scenes guidance will help nudge people toward their innate strengths without explicitly casting them into a specific role or title. Indirect coaching is a great method in situations where you're consulting a power user or a developer who doesn't have formal authority but still has a great opportunity to influence team dynamics.

Identifying the Right Mix of Leadership, Management, and Frontline Producers

Project teams must go beyond the visionary-integrator dynamic. Successful transformations require the right mix of leadership, management, and frontline producers at all levels.

Leadership Roles

Effective transformation teams include several key leadership positions:

- Executive sponsor: Secures resources, removes organizational barriers, and maintains strategic alignment
- Visionary: Provides creative direction and connects to strategic objectives
- Integrator: Ensures systematic execution and maintains focus
- Change champion: Advocates for change across the organization

Management Roles

Management roles provide the connective tissue that translates leadership direction into actionable tasks:

- Project/program manager: Coordinates activities, calendars, resources, and timeline
- Workstream leads: Manage specific sets of requirements within the implementation
- Technical lead: Oversees technical architecture, data, engineering, and systems integration
- Change management lead: Coordinates organizational adaptation to new systems or processes.

Frontline Producer Roles

Producers deliver the day-to-day task completion activities around a transformation:

- Business analysts: Translate business needs into requirements documents
- Technical implementation specialists: Configure and customize new systems
- Subject matter experts: Provide domain-specific knowledge about a system or process
- Trainers and documenters: Create materials to support training and adoption

The right mix of these roles depends on project complexity, organizational culture, and implementation approach. Outside of the visionary and integrator, in smaller engagements, one individual may have multiple roles. However, neglecting any of these layers typically leads to predictable problems:

- Missing leadership leads to lack of direction, resources, or organizational commitment.
- Missing management leads to coordination gaps and inefficient resource utilization.
- Missing producers leads to implementation delays, quality issues, and adoption challenges.

The integrator serves as the central coordination point for all roles involved in the transformation. Thus, making their role very important in the overall engagement. Marketing, sales, customer service, engineering, finance, and accounting teams should all report to the integrator within the project structure, regardless of standard organizational hierarchies. This project-specific reporting structure may vary between initiatives, allowing for customized team composition based on each transformation's unique requirements.

Balancing Technical and Non-Technical Expertise

Successful implementation teams are composed of a blend of technical and business skills. Technical-only teams build technically accurate but business-inaccurate systems that may not be usable due to an engineer-first approach. Business-only teams build unrealistic expectations of what is possible, or don't grasp the complexity involved due to technical constraints.

The technical team members bring critical competencies to digital transformation:

- In-depth knowledge of system capabilities, limitations, and upgrade paths
- Implementation competencies for designing and customizing data-first solutions
- Technology architecture designs that can ensure stability and scalability
- Security and compliance proficiency to safeguard organizational assets
- Identification of technical indebtedness before problems begin to occur

But the non-technical team members give equally valuable inputs:

- Business process knowledge that indicates how work really occurs

- Exception handling and how intricate departmental processes intertwine
- Customer or regulatory requirements that must be considered
- Strategic alignment with overall organizational goals that stretch beyond technology
- Change management awareness and personnel concerns for successful adoption
- Understanding of ROI and how investments will repay with business value

Counting only on IT to lead digital transformation normally generates technically sound systems that don't address all of the business requirements. Engineers can prioritize function over usability, solutions can miss strategic alignment, and human adoption factors are generally lightly considered. Conversely, when business units take the lead but without technical coordination, their requirements can be unrealistic and do not consider integration complexity and sustainability over time. If you are lacking technical expertise in a project, data constraints might not be evident until late in the process, triggering expensive redesign. Effective digital transformation calls for a balanced equilibrium between business and technical leadership, delivering both innovation and practicality in execution.

There are some team members who act as translators between the business point of view and technical point of view. These individuals, who understand both worlds, are essential to ensuring consistency between technical possibility and business outcomes. It is important to identify these skills or lack thereof within a team early on to ensure any gaps can be planned for and filled.

Including Diverse Experience Levels: Experts vs. Newcomers

Another crucial balance involves experience levels within the organization. Teams composed entirely of seasoned professionals may perpetuate legacy thinking. Teams of only newcomers may miss important historical context, organizational differentiators, or major client-driven requirements.

One may wonder: "Why not just hire a new team, or bring in outside consultants, to work on this?" Or, "Why not rely on my most experienced staff?" The optimal solution takes advantage of both, tapping into the knowledge of battle-hardened veterans and taking advantage of new employees' fresh eyes to devise the best digital transformation strategy

The most effective teams include:

Experts who understand:

- Historical processes and "why things are the way they are"
- Institutional knowledge about what has been tried before and the outcomes
- Established internal relationships that facilitate execution
- Potential political pitfalls based on organizational experience

Newcomers will bring:

- Fresh perspectives unconstrained by "how we've always done it"
- Current industry best practices from other organizations or competitors
- Challenges to accepted limitations or status quo
- Energy and enthusiasm for change
- Willingness to contribute, offering meaningful value to the organization

Experienced professionals possess deep knowledge of existing systems, past successes and failures, and the day-to-day realities of the business. Their existing relationships, wisdom, and risk sensitivity help them navigate any difficulties and lead to fact-based decisions.

New members, on the other hand, bring new ideas, challenge old practices, and introduce new best practices. Their energy, curiosity, and up-to-date technical knowledge many times drive innovation and keep the organization competitive.

Relying solely on experienced staff can lead to resistance to change, reliance on inefficient workarounds, and resistance to new approaches to doing things. On the other hand, relying solely on new contributors dismisses valuable history and operational concerns.

When these groups work together, tenured professionals provide the context needed for practical solutions while newcomers push boundaries and introduce innovation.

Organizations should look beyond traditional job titles when identifying potential team members. The most valuable contributors, no matter the tenure, are often the natural disruptors within the company. You are looking for individuals who consistently seek innovative approaches to challenges. These are the people who develop unconventional solutions and workarounds, sometimes creating friction with technology departments by independently adopting tools to solve immediate problems. While they may occasionally circumvent standard protocols, they represent exactly the kind of creative thinking that organizations should recognize and strategically empower within the digital transformation process.

Reflection Questions

1. In your current or recent transformation initiatives, who filled the visionary and integrator roles? Are these roles explicitly defined?
2. Are you currently balancing the creative energy needed for innovation with the execution discipline required for implementation? In which direction do you lean more heavily?
3. What processes exist in your organization for channeling new ideas during implementation? Are they working effectively? Are the majority of them being tabled for future phases?

4. How do you currently select implementation team members? Is this approach based on skills or roles within a corporate structure?
5. How well do your teams integrate technical and business perspectives? What mechanisms facilitate the integration? Who within your organization excels at speaking both technical and business languages?
6. What is your approach to developing leaders who may not naturally fit their assigned roles in transformation initiatives?
7. Are you leveraging both expert knowledge and newcomer perspectives in your implementation teams? Which way does the team lean?
8. Think about a recent project failure or challenge. Could it be attributed to an imbalance in the visionary-integrator dynamic? What specific changes might have produced a better outcome?

CHAPTER 7

Writing Great Requirements

"If you don't know where you are going, you'll end up someplace else."
—Yogi Berra

"I t looks perfect on paper. This is exactly what we need."

The project sponsor was beaming as she examined the proposal from the software vendor. The presentation and demos had been flawless—dazzling charts, seamless workflows, and seemingly every capability the executive team had mentioned during the initial discussions.

Six months later, reality had sunk in. The implementation team was overwhelmed with change requests. Users were finding fundamental holes in functionality. The vendor was discussing "custom development" at eye-watering hourly rates.

"This system was supposed to be configurable," the project sponsor lamented at an emergency steering committee. "We selected a market-leading solution specifically to minimize custom development."

"We're doing what we can," the implementation leader said, "but many of these things weren't in the requirements from the beginning. They're core to the way our business works, but they're not out-of-the-box functionality in any system."

"Weeks were spent discussing what we needed," the sponsor continued.

"We only covered what you had requested in the RFP," the implementation lead tactfully corrected. "But we never validated and documented how your business actually operates."

The difference between success and failure in digital transformation projects hinges on one thing that's too frequently performed poorly: full requirements traceability and quality. Not code quality. Not the experience of the people building it. Not even how well the software functions. All of those come into play only after the fate of the project has largely been sealed by just how well the organization has clarified what the system must do and why.

Requirements are the DNA of your implementation. They encode every aspect of the solution and determine how it will function. When that genetic code contains errors or gaps, no amount of excellent execution can produce the desired result. Unfortunately, many organizations treat requirements as a checklist rather than the foundational blueprint they truly are.

Describing Requirements in Project Communication

When we talk about projects, a fantastic psychological phenomenon takes place. When individuals utilize words to explain requirements, needs, or objectives, the mind converts those words into pictures. However, there is no means of guaranteeing that the image that one individual is seeing is identical to that which another person sees.

We all think in imagery, but mental pictures can be vastly different depending on our individual experiences, professional backgrounds, and the mental constructs we've built up over time. Two departments in a company may use the same language to explain a requirement, yet the image of what they're truly anticipating can be worlds apart.

Consider a simple example: When a marketing director tells you they require "real-time customer data," they may be envisioning colorful dashboards displaying customer sentiment and engagement metrics.

The IT director who hears the same request may be thinking of data diagrams displaying flows between databases. They are both using the same words but envisioning fundamentally different visible results.

This mental disconnect is the reason so many projects fail even with apparently well-defined requirements documents. The misalignment does not reveal itself until deliverables are shown and stakeholders understand their mental pictures were different from the start. This creates an expectation gap that ruins countless projects and squanders billions of budget dollars.

The acknowledgment of this psychological fact is the initial step to developing truly aligned project requirements.

Well-written requirements serve as an attempt to deepen the understanding between business needs and technical solutions. When requirements are poorly constructed, both sides suffer:

Business teams experience:

- Missing functionality that requires manual workarounds, resulting in increased labor costs and error-prone operations
- Interfaces that do not match workflow demands, causing user frustration and resistance to adoption
- Reports that do not provide required information, preventing data-driven decision making and strategic planning

Technical teams face ongoing challenges:

- Constant addition of new priorities, causing team burnout and eroding trust in leadership
- Re-work and scope alteration, leading to budget overruns and missed deadlines
- Difficulty estimating effort accurately, resulting in unrealistic promises to stakeholders
- Architecture designed without good information creating system instability

The cost of poor requirements extends far beyond the initial project. After the system has been launched, organizations continue to suffer from:

- Decreased operational efficiency: For example, an ERP system of a manufacturing company that has no capability to handle specialized inventory management, forcing employees to maintain parallel spreadsheets
- Increased total cost of ownership: Such as a retailer whose off-the-shelf CRM needed so many customizations that upgrade costs grew exponentially every year and vendor support was rendered ineffective
- User resistance and adoption challenges: For example, physicians not wanting to use a new patient management system since it will take twice as many clicks to perform common functions
- Loss of competitive advantage: As witnessed when a logistic firm's new system normalized their innovative cross-docking operation, removing what had been their primary market differentiator
- Damaged relationships between business and IT: Creating long-term tension that undermines future projects, such as at a financial services firm where a failed client management platform led to years' worth of IT budget battles

Case Study:
Technology's Greatest Point of Failure

Take the case of a medical device company implementing a new ERP system to replace their legacy system, which was aging. In early requirements meetings, the production staff repeatedly documented in the RFP that they wanted to be able to "track inventory effectively" and "view accurate stock levels." These requirements were diligently captured by the implementation team, and the vendor made sure their system could accommodate typical inventory management.

The project sailed smoothly through design and early implementation. Executives were happy with the status reports, and the rollout was two weeks ahead of schedule and slightly under budget, an unusual success in the world of ERP projects.

Then came user acceptance testing.

"This can't be right," exclaimed the Quality Assurance Manager in a UAT session, her voice climbing in dismay as she gazed at the screen. "Where do I put the lot number? Where's the expiration date? How do I track the serialized components?"

The room fell quiet. The implementation consultant appeared puzzled. "You didn't specify lot control or serialization in the requirements," he stated, bringing up the requirements document on his PC.

"Of course we require lot control," the QA Manager responded, her face changing from a look of puzzlement to one of shock. "We're a medical device firm. We have to be able to trace every single component through the entire production process. If we ever have to recall, we have to know precisely what devices we have and what components from what supplier lots. It's not only a business requirement, it's regulated. How could you have possibly missed that?"

The Production Manager agreed. "That's why we stressed 'effective' inventory tracking. Lot control is so central to our business that we thought that any inventory system would include it. It would be like buying an automobile assuming that you would want wheels."

Six months into implementation, with go-live just weeks away, the project hit a massive, unforeseen obstacle. Adding lot control and serialization capabilities required:

- Completely redesigning the inventory data structures to include lot numbers, serial numbers, and expiration dates
- Modifying more than 20 screens across purchasing, receiving, production, and shipping modules to include the fields

- Updating every inventory report to include traceability information
- Creating entirely new quality control and receiving workflows
- Developing custom validation logic to enforce entry requirements
- Reconfiguring every integration with upstream and downstream systems

The ripple effects were devastating. The go-live date was pushed back by four months. The budget increased by $800K, a 40% overrun. The implementation team had to be almost completely rebuilt as key consultants rolled off to other projects due to scheduling issues. Most damaging of all was the loss of confidence: executives questioned the competence of both the department and the vendor, while end users became increasingly skeptical that the system would ever meet their needs.

"How could we miss something so fundamental?" the CIO asked during a particularly tense steering committee meeting.

The answer was painfully simple: no one had asked the right level of detailed questions when building the RFP or during the design processes. The business users assumed lot control was standard system functionality and it didn't need to be explicitly stated. The implementation team didn't have enough industry knowledge to know what effective inventory tracking meant in a medical device company. The vendor's sales team had focused on showcasing dashboards and reporting rather than validating core operational requirements.

This wasn't a technical failure. It wasn't even a failure of the software product. It was a requirements failure. The company hadn't properly defined what track inventory actually meant in their specific regulatory and operational context. A few days of detailed requirements workshops could have prevented months of delay and hundreds of thousands in cost overruns.

The Anatomy of Successful Requirements

Requirements quality strongly affects the implementation success, but most organizations find it difficult to create concise, testable requirements. The distinction between good and bad requirements is the language that is used.

Consider these examples:

- Poorly-written requirement: "The system shall be user-friendly and have the capability of entering customer data easily."
  - These criteria have no objective measures of success. What is easy or user-friendly is purely subjective. There is no way to make an objective determination of whether or not the implementation satisfies these vague requirements.
- Well-written requirement: "The customer data entry screen shall permit a trained user to input a new residential customer record, name, contact details, and service address in under 60 seconds with not more than 5 tab/field navigations. The system shall perform real-time address validation against USPS standards and shall prevent the duplication of customers based on matching full name and address."
  - This requirement gives concise, testable specifications. It defines the user type, the task, the desired level of performance, and what needs to be validated. Implementation teams can design a solution based on this specification, and tests can clearly be written to determine whether the requirement has been fulfilled.
- Poorly-written requirement: "The system shall provide detailed sales reports."
  - This statement does not specify what sales reports are required, what information they must contain, who requires them, or for what their purpose is.
- Well-written requirement: "The system will generate a monthly revenue forecast report with projected revenue by product category, territory, and sales rep for the upcoming three

concurrent months. The report must be exportable to Excel, be automatically emailed to regional managers on the 1st of every month, and include comparative data from the previous year's actual performance in the same row. The forecast calculation must include both pipeline opportunities weighted by probability as well as recurring revenue from installed contracts."

    o   This is not a requirement that has much area for interpretation. It specifies content, format, distribution, timing, and method of calculation, and gives developers and configurators clear directions on what to design at the next level.

A Requirements Alignment Framework

The previous example of a medical device ERP implementation illustrates how requirements may be documented using differing levels of complexity and sophistication. It is here that a questioning framework proves helpful in creating structured questions around requirements-gathering. At a fundamental level, ask the team this question: "What requirement are we trying to solve?" Being explicit about the words that are used when documenting requirements is critical for setting the overall objective. There are truly only a handful of questions you should ask and understand when you write the requirements. This same language should be the guide used to govern the final testing and go-live of the requirements. It guides the traceability through every requirement. Notice the questions posed below:

- Surveillance or visibility: Are we solving for the problem of being able to view and monitor significant information? In this last example, it was as simple as viewing lot tracking data in the system. This is the basic level at which visibility into significant information starts.
- Performance or reporting: Are we solving for the problem of getting insights from the data? The team anticipated using the surveillance data. In our case, this meant planning product

lifecycle management and recall planning using lot data. This phase transforms data into useful information to answer the question, "Are we meeting metrics or organizational measures?"

- Excellence and process: Are we ensuring every person in the company follows the same process every time? The medical device manufacturer required all inventory transactions to record lot and serialization information by everyone every time, enabling excellence and full traceability across operations.

- Workflow automation: Does the requirement or solution intend to completely automate end-to-end processes? In our example, there may be a requirement to automate the whole chain of custody with both vendors and distributors, developing an end-to-end tracking mechanism along the supply chain.

Only after it is understood what the requirement intends to provide (visibility, reporting, process, or automation) can the requirements be written. Once they are written to this level of specificity the roadmap to implementation can be drawn.

It is essential as a part of this process to know where your organization is today and where you hope to end up with respect to these questions. If an organization is currently struggling with surveillance and departments do not have access to good data, visible to everyone who needs it, then the journey to full automation is going to be daunting. Each progressive step forward, visibility to reporting, reporting to excellence, etc., increases the complexity, timeframe, and costs of a requirement. An organization attempting to jump from simple surveillance to complete automation without intermediary steps will certainly undergo extreme implementation hardships. That's the reason why, even though departments will often say something along the lines of "We don't want you to examine what we have today since that's not what we want in the future," you need to record the current state, nonetheless. Recording the current state offers great insight into how a company views its processes and what mental models the department leaders are working from. That as-is information is the basis for the ability to explain the to-be state with clarity. A full understanding of the gap will allow you to write a requirement in a way that clearly states where you intend to be. Examine the example below from an actual RFP we have received and the alternative approach using these questions.

Example requirement from an RFP: "The system shall allow for customer relationship tracking."

Revised using this method: "The system shall provide visibility to the sales team to their own customers and visibility to the relationships with other accounts, contacts, or prospects."

Note: This change in language clarifies that the goal of the requirement is to provide visibility. (surveillance) to data in the system. There is no mention of automated data capture by team members (automation) or even a standardized process for how relationships would be built or categorized in the new system (excellence). Once an as-is analysis is performed, stakeholders can understand where the company is today and what it will take to get them to the surveillance stage.

Visualizing Early: Closing the Mental Model Gap

As we mentioned above, language forms images that are different from person to person. In requirements documentation, the disparity is greatest when specific words have very different meanings to members of the technical team.

Consider this example requirement from an actual RFP: "The system shall automate the management of inventories."

To a warehouse manager, "automate the management" would imply a user-driven process by which they review and approve system-recommended levels of inventory. To a technical engineer, the same term may bring to mind a purely backend process with no user interface whatsoever, where algorithms calculate and perform adjustments without any user intervention. To the CFO, "management of inventories" could mean reconciliation of financials surrounding inventory. All of these meet the literal expectation but require solutions that are extremely different.

This potential for ambiguity is the reason that written requirements by themselves, while extremely important, are inadequate. We need early visualization through wireframes, prototypes, and system demos to achieve alignment of mental models. When a stakeholder is able to point to a screen and say, "Yes, that's what I had in mind," or, "No, that's not at all what I had in mind," we create the chance to repair misalignments prior to expensive reimplementation cycles.

Consider a requirements workshop where, rather than simply writing that "the system shall provide visibility into customer interactions," the facilitator presents samples of several CRM dashboards showing emails, call queues, etc. This tangible visual cue enables stakeholders to define precisely what information they require, how it is to be presented, and what immediate gaps there may be.

By presenting visual elements early and throughout the requirements process, teams can ensure that they're designing toward a common vision instead of individual mental images that don't surface until it's too late to alter course.

The Hierarchy of Requirements

Not all requirements are created equal. Indeed, assuming that they are is a surefire way to disaster, and an endless list of features with no priority or link to business value.

Good requirements form a hierarchical framework that connects business strategy to detail. The Level 1-6 requirements framework presented in Chapter 3 constitutes such a hierarchical approach. As you go from Level 1 to Level 6, the requirements get progressively more detailed and progressively more technical. At the same time, the audience transitions from business stakeholders to technical implementers.

Most requirements fail because organizations:

- Cut directly to the Level 5-6 technical detail without providing decision and value-level context:
  - *Example*: A retailer jumped straight into rigorous UI specifications for their inventory management screens without nailing down up front what business goals the system needed to fulfill. The result? A beautifully designed system that tracked the wrong metrics, requiring a costly redesign after launch.
- Stop at Level 1-3 strategy level without converting it into actionable requirements:
  - *Example*: One vendor's EHR implementation got mired in "improve patient onboarding experience" without defining what particular workflows were to be improved. Development teams made assumptions on their way into creating sets of features, which resulted in a 40% adoption rate due to clinicians realizing that the system failed to address their real pain points. Had they seen the technical design prior to build ,the clinicians could have spotted potential discrepancies.
- Circumvent critical gates in the process, causing a disconnect between strategy and action:
  - *Example*: An insurance company developed detailed technical specifications (Level 6) and had upper-level strategic objectives (Level 1) but was missing the essential middle levels where business processes ought to have been specified. The system that resulted exactly met the technical requirements but could not support key business functions, necessitating $2.3M in post-implementation rework.

Within each requirement, effort and value must be assessed for each of the requirements. A valuable stakeholder alignment tool is the prioritization quadrant chart that plots requirements against:

1. Business value (low to high): The direct effect on operating efficiency, revenue growth, cost reduction measures, or strategic goals
2. Implementation effort (low to high): The effort, costs, resources, and complexity required to deliver

This sample graph creates a visual four quadrants for the team to see at a glance:

1. Quick wins (low effort, high value): Become priorities for implementation with rapid ROI
2. Strategic projects (high value, high effort): Major projects that require careful planning but lead to high payoff
3. Fill-in work (low value, low effort): Easy enhancements that can be included when resources allow
4. Avoid or minimize (low value, high effort): Requirements that must be challenged or deprioritized altogether

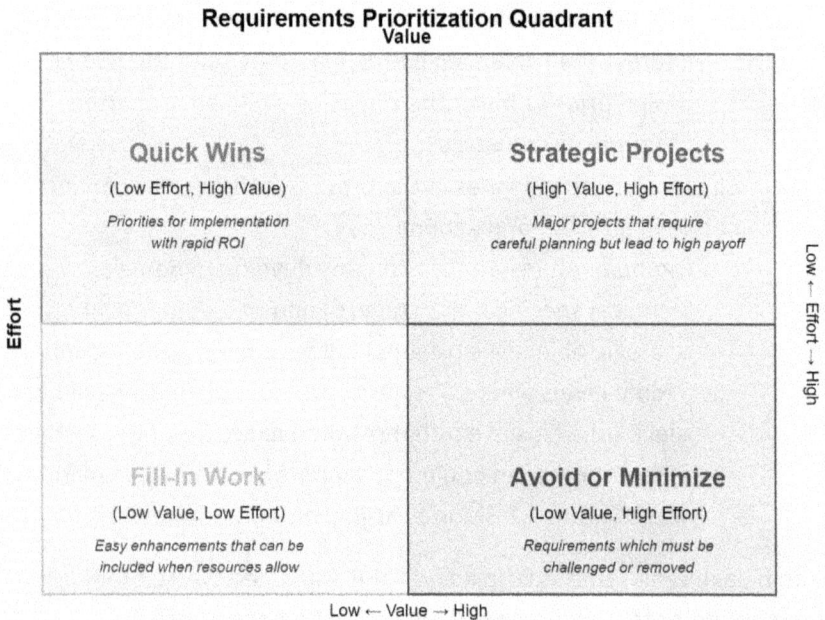

Requirements Prioritization Quadrant

Value

Quick Wins	**Strategic Projects**
(Low Effort, High Value)	(High Value, High Effort)
Priorities for implementation with rapid ROI	*Major projects that require careful planning but lead to high payoff*
Fill-In Work	**Avoid or Minimize**
(Low Value, Low Effort)	(Low Value, High Effort)
Easy enhancements that can be included when resources allow	*Requirements which must be challenged or removed*

Effort

Low ← Effort → High

Low ← Value → High

This model compels stakeholders to decide on trade-offs and reach consensus on priorities. When, for instance, a full set of ERP requirements was plotted for a major steel manufacturing firm, executives who were demanding their own priority lists reached consensus on phased implementation after clearly visualizing the value/effort distribution in the graph. It became clear that their preconceived priorities were off-base based on the value to the organization and the effort each would demand.

Another critical component is understanding the requirement dependency. Certain requirements are foundations upon which others are built and must be implemented prior to those that depend on them. Building a dependency tree avoids the frequent trap of trying to implement dependent features prior to the foundations upon which they rest, only to find it necessary to rework and exceed budgets. For example, at one telecommunications company, the team attempted to apply sophisticated customer segmentation capabilities prior to being able to solidify customer hierarchy data. The outcome was several months of delay since they needed to go back and reconstruct the data hierarchy first before continuing the work on segmentation that they had begun.

The Illusion of "Best-of-Breed" Software: Why COTS Solutions Are Not Magic Pills for Simplified Requirements Definition

Organizations often seek shortcuts around thorough requirements definition because the effort to define requirements fully is hard work. It requires diligence and true organizational alignment. They do this by selecting commercial off-the-shelf (COTS) or best-of-breed industry-specific software. The thinking goes:

> "The system will tell us how to run best practice processes; since we don't have standard operating procedures today, this will make implementation easier."

> "We can avoid the hard work of defining requirements by adopting industry best practices."

"This software is designed specifically for our industry, so it must be a good fit."

This logic is rarely proven to be true and many times a COTS is just the easy button for lazy decision makers. Although vertical market software might embody predefined best practices, it is no replacement for specifying and refining your own business processes. Businesses that haven't bothered to document their own requirements and their existing process are making a catastrophic mistake. They foolishly hope software will impose discipline they themselves have not bothered to define, basically trying to use technology to solve for organizational discipline. This almost always results in failed implementations or achieves only a fraction of its possible benefit. A COTS solution will many times shoehorn your company into the workflows created by an external vendor, not the workflows that are most valuable to the company.

Take this example: We have found that many construction contractors, for instance, are captivated with the idea that an ERP package specific to their industry will be a turnkey operation. Consider a well-known commercial builder, a mid-sized contractor that believed an industry-specific construction ERP was the overnight cure for their project management and budgeting challenges. "The software understands construction contracting and job costing more than we do," bragged their director of operations when questioned about bypassing process and requirements documentation. Six months after implementation, they discovered the bitter reality of this shortcut. Their proprietary process for managing change orders, a key competitive differentiator, could not be configured within the system's standardized workflow. We find the majority of construction contractors make the same mistake, mesmerized by the notion that an ERP package aimed at their industry will be an easy solution.

Vendors sell the illusion of out-of-the-box efficiency, compliance, and automation. What they fail to mention is that the systems are designed with generic processes that might not mirror the manner in which a specific company does business.

Following are the common pitfalls of this approach.

Lack of Fit Due to Undefined Internal Processes

If you haven't captured the way your business really works, you have no foundation for making the determination whether or not a COTS software application is going to work for you. Firms that won't document their processes and detailed requirements are flying blind to an unknown destination and expecting to land in their dream location. A firm cannot assume that pricey software will intuitively understand the business and its processes better than they do. They are constructing a house with no blueprints and then blaming the contractor when the kitchen is in the wrong location. That basic mismatch results in implementation fiascos that do not surface until it is too late.

Example: A pharmaceutical distribution company selected a highly rated pharmaceutical warehouse management system after being impressed with a vendor demonstration of such features as picking, zone management, and automatic put-away. Six months into the implementation, they discovered a fundamental problem. The embedded workflows within the system were designed for high-volume distribution of commodity pharmaceuticals, but their warehouse was designed and built around a small-batch specialty pharmaceutical business that required high-touch quality inspection. This specialty line within their business made up more than 60% of their profit and was their competitive differentiator.

The company was presented with a bitter choice to either reengineer their entire warehouse operations (abandoning much of their competitive forte of specialty fulfillment at speed) or invest millions customizing the software. The project languished for months as executives battled over direction, all while still paying for consultants, licenses, and infrastructure.

This expensive pause might have been prevented if they had only documented how their warehouse really worked and what a system needed to support before they chose the software. Had they done that, they would have been able to recognize this mismatch at selection time when it would have been relatively inexpensive to change direction.

False Expectation of Plug-and-Play Implementation

Most organizations are guilty of expecting that they can simply install industry-specific software and it will work its magic in solving their problems of operation. The dream is that you bypass requirements-gathering, design, and development and jump directly to implementation. But this brings up two critical questions:

- How do you know what to train users on when you haven't done the up-front requirements work?
- How do you test when you don't even know what success looks like for your particular business?

Without requirements, training becomes general software training instead of training specific to a particular business and what company procedures are supported by the system. Testing becomes a way to check the boxes, proving you can perform generic functions rather than running specific war game scenarios that will occur in real-life business operations. This type of approach is an abdication of process ownership and a denial that implementation will involve hard work.

Example: A regional insurance company selected a claims processing system without understanding it was specifically designed for property and casualty insurers. The executive team was sold on the promise that the system embodied insurance industry best practices and would modernize their operations without requiring them to redesign their processes.

When implementation began, the reality hit hard. Despite being insurance industry-specific, the system still required:

- Defining over 200 configuration parameters and adding user-defined fields
- Making decisions about approval workflows and authority levels that were undefined

- Specifying document management rules and retention policies never previously discussed
- Establishing integration points with underwriting and billing systems
- Creating custom correspondence templates because of specific client requirements
- Defining user roles and security profiles only to find that the software could not support the roles and hierarchy in alignment with the company structure

Every option configured needed detailed knowledge of their existing processes, future state desired, and trade-offs between standard functionality and business needs. Lacking this foundation, the implementation team was faced with the same level of design meetings and the struggle to make important business decisions in the context of an even shorter project timeline and with limited flexibility within a rigid COTS solution.

No software, regardless of how industry-specific, can substitute for the work of process design and optimization. Even out-of-the-box solutions demand considerable configuration, data migration, and change management effort.

There is a dirty little secret in the business solutions software marketplace that vendors do not usually discuss. Most so-called commercial off-the-shelf (COTS) products are vaporware, software that is sold and marketed prior to being in a finished form. For software firms, the two biggest hurdles to clear are getting that first reference customer and using it to fund the actual development. Some providers have found that a convenient way out is to offer a product as if it already exists and charge customers implementation fees to cover the cost of its creation.

This is more common than most business leaders know. The vendor presents slick mockups and over-scripted demonstrations that feature theoretical capabilities rather than working software. They might have a platform or framework in the middle, but the industry-specific capabilities they are selling are very much PowerPoint slides and project plans for future development. They will build the shell of an industry-specific product and then custom build vertical features as a part of implementation without fully disclosing the level of development needed to get the features available for the inaugural client.

Look at what occurred at a large auction house that was looking to update its business. They reviewed a number of vendors and chose one that seemed to possess the ideal blend of ERP, online portal, and specialized auction management functionality. The demos were good, and the references (though limited) were acceptable. The auction house executed a large contract, glad to have discovered an existing solution—or one they thought existed—that would not need huge amounts of customization.

Their relief was temporary, however. Instead of jumping straight to implementation, they were surprised that the vendor was doing complete requirements-gathering activities in excruciating detail and asking questions as if the system did not exist. When they asked for demonstrations of how particular functionality was present in the real system, they received vague answers. They finally discovered the truth: a group of developers was desperately creating the auction-specific functionality from the ground up based on the requirements being specified in those same workshops.

"What we believed was a turnkey product ended up being worse than beginning with a platform solution," their CIO later elaborated. "We paid a premium for what we assumed was tested functionality, only to be the guinea pig for their inaugural live auction industry implementation. We weren't configuration customers; we were unintentional R&D partners."

This is a situation that is profitable for software firms who are essentially getting paid for developing their product that can be resold to future clients, but is greatly unfavorable for initial clients because they are left with:

- Timeline uncertainty: Schedules for development are highly uncertain, especially for never-before-built functionality.

- Quality issues: More recent software inevitably has more bugs and design flaws than software that had been through the fire of multiple implementations.
- Higher implementation risk: There is no tested-and-proven implementation approach because the project is an experiment regarding both processes and software.
- Surprise customization expenses: "Standard" features that were guaranteed still need costly custom development. This process will not be the last time a vendor does this. It's a matter of time before the expensive change orders begin to come in.

Vendors can be vetted for this by demanding rigorous proof during the purchasing process: exact demonstrations of specific functionality (not generic workflows), access to the real system features to test rather than mock-ups, and direct conversations with referring customers using the same modules being purchased in your specific vertical industry. Most of all, ask the direct question: "How many customers are running this specific module in production today?" and "Are you building or configuring anything specific for our company, and if so, what?" The answer, or any hesitation, will be revealing.

Rigid Workflows That Don't Adapt to Business Needs

COTS solutions come with many pre-defined processes that may not match your organization's unique operational requirements. When mismatches occur, organizations often find themselves contorting their operations to fit the software rather than having technology support their business. The major risk is that your business will quickly outgrow the system as it proves inflexible, unable to adapt when you change direction, add new offerings, or expand into adjacent markets.

Example: A construction consulting company that provides professional services for an industry-specific project management and accounting software went with a COTS solution. The software was robust in milestone-based billing, which was well-suited to their primary business model. Implementation was smooth, user adoption was high, and the system delivered value immediately without customization.

Two years later, the company moved into facilities management consulting, a practice that billed based on time and materials (T&M) instead of billing by milestone. To their astonishment, their "ideal" operations management system could not manage this billing mechanism and had no functionality for time tracking. The system architecture had been designed with the presupposition that all projects adhered to a predictable stage-gate process prior to billing.

"We did everything we could to try to make it work," their PMO director clarified. "But the system just wasn't up to capturing consultants' time or producing the activity reports our T&M clients needed."

The firm ended up putting in a stand-alone time and billing system for their T&M business. This generated a technical nightmare: two systems, duplicate customer and project data entry, reconciliation issues for invoices, and fragmented reporting. Even worse, they needed to have project managers using two different systems, which produced inefficiency and confusion.

"What appeared to be an effective workaround solution turned out to be a tremendous drag on our ability to grow," the director went on. "We're really operating two businesses with different systems, different processes, and different metrics. Our COTS solution that we thought would save us money in the past now costs us more in operational inefficiency than full custom development would have."

Where COTS applications impose rigid processes, companies are faced with a choice: conform to the system's process (maybe at the expense of competitive differentiators) or invest in customizations that nullify the cost benefit of buying an off-the-shelf application. Worse, most companies find their growth stunted by technology that cannot support expansion into new services or markets. The idea of new business growth becomes impossibly expensive since they are held back by technological constraints, necessitating yet another digital transformation initiative, and thus the cycle continues.

When COTS Solutions Are Appropriate

In spite of these drawbacks, COTS products may be suitable in certain situations:

- When your activities match the practices of the industry, and you are ready to follow the best norms with no plans for expansion into new markets or varying lines of business
- For non-core, back-office functions where competitive differentiation is not essential (i.e call routing, PBX, or VOIP systems, as long as they can support the needs of the organization)
- When the urgency of execution trumps the requirement for flawless alignment or the potential rework that will be required in a later phase
- For small and medium-sized organizations with fewer IT resources and budgets to support enterprise-grade systems or platforms
- When the required functionality is extremely well specified, the system checks all the boxes for requirements, and the business is unlikely to change

The best option for most businesses is to choose a platform solution that can adapt over an inflexible COTS product. Platforms such as Salesforce, Microsoft Dynamics, SAP, or ServiceNow provide tremendous benefits:

- Core back office functionality can be utilized directly out of the box to manage generic processes where specificity is not required.
- Options for customization enable you to shape the system according to your own needs while still making use of core functionality.

- Huge ecosystems of ISVs (independent software vendors) offer pre-existing solutions to particular industry requirements that bolt on to the overall platform.

Take local government deployments, for instance. A city may use Microsoft Dynamics 365 as its platform for financial management and then layer on specialized ISV solutions for things like permit management, tax collection, and citizen services, all of which extend from a common data platform but meet distinct municipal needs. They may also build their own add-ons where other vendor solutions fall short, but still do it in a way that is supported as if it was managed by a third-party vendor.

When selecting a platform and implementation partners, you should prioritize:

- Suppliers who have demonstrated stability and longevity: In the AI age, there are many vendors popping up that offer platform services without the legacy of being in such an industry.
- Technology that is aligned with your current IT setup: For example, it may be unwise to use a Microsoft Business Solutions platform when your organization runs AWS and Google Docs for core organizational daily use.
- Partners who have a specific implementation plan: For example, covering the principles laid out in this book and who have a proven track record of sticking to their plans.
- Team members who will stay constant throughout the project life cycle: This ensures that you are not left with constant turnover and reeducation processes.

A Word of Caution About Implementation Partner-Developed Solutions

An extremely risky approach during implementation is leveraging pre-built software solutions developed by your implementation partner. While this may be tempting, providing faster implementation with pre-built needed functionality, it sets you up for dangerous dependencies. If things don't work out with your implementation partner, you're left with undesirable choices: remain with a partner who does not meet your needs so you can keep the existing software in place, lose support for critical functionality, and switch partners, or pay astronomical fees to switch to an alternate partner and rework software functionality. As a rule of thumb, choose software vendors for software and system integrators for implementations. This separation provides healthy checks and balances and maintains your flexibility going forward.

The retail price of COTS software is usually just the tip of the iceberg. The total cost encompasses many hidden charges that can exponentially add to the investment and timeline, undermining the very benefits they claim to provide.

Example: A mid-market manufacturer chose an industry-specific ERP package with an initial license price of $400K, well below the estimated $1.2 million of a more configurable, general-purpose solution from both SAP and Microsoft. The choice appeared to be a victory for the company budget.

Two years later, a full accounting of the project told a different story:

- $750,000 in customization costs to adapt the system to their unique production scheduling requirements and provide the reporting and dashboards needed to operate
- $200,000 for integration with existing CRM systems so that quoting and potential sales could be included in the production planning process
- $300,000 in internal labor costs for data cleansing and migration because these tasks were outside the scope of the software vendor
- $250,000 in productivity losses because software features were not available during transition and needed to be built as part of a phase two engagement

- $180,000 annually in new hiring internally for new development and support resources who understood both the business and the customized system

What initially seemed like a $400K investment escalated to almost $2 million in implementation costs alone, with maintenance costs continuing to far surpass initial expectations. In the meantime, the allegedly more costly general-purpose solution may have offered greater long-term value by delivering greater flexibility and fewer customization requirements. It could well have provided more scalability because of the features that were not implemented in phase one that could be scaled into as the organization evolved.

The Acquisition Risk:
When Your Software Is Sold Off

Perhaps the most unpredictable cost factor in COTS offerings that most do not consider is the risk of vendor acquisition. This is a factor that's impossible to fully account for during selection. Consider the recent acquisition of VMware by Broadcom, for instance, which caught enterprise IT organizations worldwide by surprise.

Having finished the $69 billion acquisition, Broadcom wasted no time in taking its strategy to the extreme. It aggressively raised license costs, up to 200-400% in certain instances, while also abandoning perpetual licensing models in favor of subscriptions. Most troubling, it intentionally reduced its customer base by targeting only the largest enterprise customers, in effect leaving behind mid-market organizations that had built their infrastructure around their platform products.

"We were forced to make an impossible decision," said the CIO at a manufacturing company. "Because of an acquisition over which we had no control, we either had to pay the astronomically higher charges or undertake an even more costly, risky migration to other solutions."

This situation plays out with eerie regularity in the enterprise software world. In choosing COTS solutions, organizations need to understand they're not merely purchasing software; they're entering into a long-term relationship with a vendor that can see its business priorities change radically. Today's customer-oriented software vendor will become tomorrow's profit-making acquisition target.

Actionable Steps for True Total Cost Analysis

1. Ask for an in-depth implementation estimate and project plan: Require complete breakdowns of every phase of implementation, such as customization, integration, data migration, and training.
2. Calculate internal resource commitments: Measure the hours your staff will devote to the project and multiply by relevant hourly costs, including things like opportunity costs.
3. Conduct reference checks for cost surprises: Ask current customers for cost surprises they experienced during implementation.
4. Establish transition cost models: Project the losses in productivity during cutover phases when systems and users are transitioning.
5. Estimate the 5-year support costs: Not just vendor maintenance charges but also in-house support personnel and upgrade costs.
6. Assess vendor acquisition risk: Research the software vendor's financial health, ownership, and acquisition history within their market space.
7. Calculate exit costs: Figure out what it would take to move to a different solution if the relationship sours, so you're not caught in a technical hostage crisis.
8. Capture process impact costs: Determine processes that will be affected or need workarounds and estimate their impact on business operations.

Most importantly, have specific team members question every cost assumption, playing devil's advocate to root out hidden costs before they turn into budget-breaking surprises. The aim is not to eschew solutions altogether, but to go into these agreements with eyes wide open to the overall fiscal impact outside of the up-front cost.

Case Study: The Hidden Requirements

"Go-live is eight weeks away. We are on schedule and on budget," the Project Lead reported at the executive steering committee meeting. A PowerPoint full of green status lights adorned the presentation slide. "The vendor has completed 92% of configuration, we've migrated all historic data, and user acceptance testing is scheduled for next week."

The worldwide shipping company had spent $12 million on a new transportation management system (TMS) to take the place of their outdated platform. The requirements phase itself lasted 18 months and delivered a detailed document outlining routing, rating, scheduling, carrier management, and all the supporting documentation. The implementation had gone smoothly for the last 24 months.

The Chief Financial Officer nodded in agreement. After two failed attempts at upgrading their technology over the last ten years, they had gotten it right at last. This time, they had done everything by the book—comprehensive requirements, experienced implementation partner, committed internal team, and iron-fisted project governance.

"Going on to the UAT timeline," the Project Lead went on, but was cut short by someone calling out from the back of the room.

"Are we going to include retention billing?" said a woman who had been sitting quietly in a corner seat.

The Project Lead broke off, looking through his papers. "Sorry, who are you?"

"Michelle Summers. I head the retention team in Singapore." She did not attend the meeting but was at headquarters for a training session and overheard someone mention the project update.

The Project Lead's face was stone-like. "Retention billing? Is it within the scope of Phase 1?"

The Chief Information Officer scowled, staring at his tablet. "I don't remember that being discussed."

Michelle's expression changed from amazement to alarm. "Retention charges produce $52 million in revenue each year. That's what we bill customers for retaining containers beyond their expiration date. Tell me that's included in the implementation."

The room fell into silence. The Project Lead nodded toward the technical expert, who rapidly scrolled through the requirements document on his laptop.

"I don't see any retention billing requirements per se," the technical expert affirmed. "There's a generic section on billing integration, but nothing on container retention calculations or specialized rate structures."

"That's impossible," Michelle exclaimed. "It's the most complicated billing routine in the whole company. We monitor thousands of containers in 43 countries, charge time against various customer-specific contracted allowances, assign variable rate structures depending on location and container type, and interface it into the bill system. How will this be handled after migration to the new system?"

The vendor representative coughed awkwardly. "Our core TMS does not include retention management. None of our previous customers in your industry has requested it. This would require extensive custom development."

The Project Lead's confidence waned as the implications sank in. The system they had designed for the last 18 months was unable to process a key revenue-earning process. The meticulously scheduled go-live date now looked out of reach.

During the tense post-mortem that ensued, they found a number of failures in their requirements process.

The initial problem was the assumption of knowledge. The logistics group assumed retention billing was core to their business and that it would be a part of any TMS system.

In the meantime, the IT department was unaware of the business importance or technical complexity of this process. And the vendor, although they were industry specialists, had never implemented this feature for another customer and never inquired. "How could we miss something generating $52 million a year?" the Chief Information Officer wondered, running his hands through his hair.

"Since we collected requirements department by department, rather than by process," explained the business analyst, referring to their guide. "The retention process reports under Finance in Asia, but the process impacts operations, customer service, and legal for all entities. We did our requirements sessions by departmental functions, so this cross-department process fell through the cracks. We did not have anyone from finance in Asia in the other department sessions."

This revealed the second breakdown: siloed requirements-gathering. No one had tracked the end-to-end billing process department by department. The operations group had discussed tracking of containers, but not retention specifically. Finance had discussed billing integration, but not the customized calculations for retention charges. The key integration points between the systems were lost because each group only thought through their piece of the process.

The third breakdown was in representing the users. The people actually doing the retention billing, primarily in Singapore, were not in the billing requirements meetings conducted primarily by the U.S. billing team.

"We've talked to the regional director," the business analyst responded curtly.

"They have not dealt directly with retention billing in fifteen years," Michelle said. "Things have changed a lot since that time."

This was how their process documentation represented ideals, rather than real day-to-day practice. The knowledge that was vital, workarounds, exception handling, and expert calculations had not been captured because the individuals who possessed this knowledge had not been included.

The implications were severe. The project team was forced to pause the implementation, formally document the end-to-end retention billing process, engage the concerned stakeholders in multiple global offices, and create a customized module for this critical function. This unexpected work increased the project timeline by four months and the budget by $1.8M.

Even more damaging was the loss of credibility. The Project Lead was removed, and IT's credibility with operations was severely eroded. "How do we know what else we might have missed?" the CEO had asked at one particularly tough meeting. It was a question that no one could answer for sure. The lesson was excruciatingly obvious: detailed requirements-gathering is a key success factor for any project. Hidden requirements always surface. The only variable is whether you'll find them through methodical analysis or during the crisis of a failed implementation. The expense of correcting them late in the project's life is exponentially greater, both in dollar terms and organizational terms.

Instead of using industry-specific software as a shortcut to avoid defining internal processes, organizations should take a disciplined approach to system selection. This ensures that regardless of the method you pursue, custom development, platform configuration, or COTS implementation, the resulting system will meet your business requirements rather than requiring you to adapt to the system's limitations.

- Fully document and analyze current processes: Prior to being able to assess any technology, fully document how your company really works, not how it's theoretically intended to work or how management thinks it works. Inefficiencies, bottlenecks, and opportunities for improvement should be uncovered by this analysis as you document critical processes. Process documentation is the measuring stick against which prospective solutions are evaluated.

- Perform a fit-gap analysis: After processes are mapped, methodically examine how prospective solutions align with your needs. The comparison should measure the extent of fit and the prominence of gaps, giving you an objective comparison instead of a personal preference for known vendors or flashy demos.

- Determine if you can adapt: There will be gaps with every solution. If you don't find any then you have not done a thorough enough job yet. For every major gap, make conscious choices regarding whether your company will accommodate the system or if you will customize the system to fit your needs. It is important to take into account technical gaps as well as gaps in organizational readiness for change and long-term strategy.

- Consider a hybrid approach: The best implementations take a mixed approach, such as applying one-off functionality where it matches requirements and creating tailored solutions for genuinely distinct processes. This strategy maintains competitive distinction yet uses proven best practices where it can.

- Think long-term about TCO and agility: Consider possible solutions not only for today but how they will scale with your business as it grows and changes. A solution that meets the requirements of today but cannot adapt to tomorrow's business creates technical debt that ultimately negates short-term savings.

The Critical Role of RFPs in System Selection

The request for proposal (RFP) process is one of the most important milestones in enterprise system selection. RFPs and RFIs (requests for information) are the usual initial steps of organizations when purchasing large software systems. They are used to document initial requirements, obtain comparable responses from vendors, and establish an objective selection process with an end result of selecting a software and platform to implement. Deceptively simple in concept, the RFP process is replete with complexity and unwarranted influence on project success.

The majority of RFPs come from one of a number of sources: internal procurement groups working from templates, IT departments writing up specs, business units establishing the operational requirements, or third-party consultants with expertise in vendor selection. Wherever they come from, the quality of this document will determine the entire implementation process. A well-crafted RFP brings in the right vendors and establishes clear expectations; a sloppy one essentially ensures project problems before selection and implementation begins.

RFP Mistakes to Avoid

Requirements definition is particularly critical when developing RFPs. A flawed RFP virtually guarantees a flawed implementation. Here are common RFP mistakes to avoid:

- Generic requirements lists

   One of the most common errors in RFP development is the wholesale copying of generic requirements lists. Organizations replicate requirements from industry templates, past projects, or consultant libraries without strict review. It leads to inflated documents filled with unnecessary specifications that bury truly essential needs. Worse, generic requirements do not capture the organization's distinctive operational modes that currently exist.

For instance, a manufacturer may have generic inventory management requirements lifted from a template but not define their particular lot tracking requirements or quality control process steps. This omission compels vendors to address dozens of inconsequential details and overlook the essential capabilities that will dictate implementation success. The outcome? Proposals that appear complete on paper, because of their size, but lack fundamental functionality in reality.

- Vague language

    RFPs are also notorious for imprecise terminology that invites confusion and sets the stage for post-selection scope issues. Adjectives like user-friendly, robust, flexible, or state-of-the-art are invoked without definition, creating an issue where vendors see what they want to see. The term "intuitive interface" implies minimal clicks to one organization, while to another it implies contextual help functions. Very different sets of functionality.

    This lack of language precision carries over to other requirements. Absent concrete metrics such as "support 500 simultaneous users with response times of less than a second" or "close month-end in four hours or less," vendors cannot properly size hardware needs or validate performance guides. When the specific requirements do finally materialize during implementation, they tend to be costly infrastructure additions or custom optimizations that were not budgeted for.

The key is precision. Substitute objective language in the requirements with qualitative descriptions that allow no room for interpretation. In communicating ambiguous requirements, detail precise workflows, screen transitions, system maps, or specific metrics instead of qualitative judgments made based on reading "friendliness."

- Missing process context

The worst of all RFP errors is listing features without identifying the business problem they solve. Requests such as "must have workflow approval" or "requires inventory tracking" are included with no background information on how these functions are integrated into the operations of the organization.

This lack of process context does not allow vendors to understand the underlying needs behind the feature requests. A CRM listing the need for "lead scoring" will have a completely different meaning to a high-volume B2C company than to a low-volume, high-touch B2B company. Without this context, vendors react with boilerplate functionality that can ultimately misalign business requirements. Worse yet, they simply respond "yes" to every requirement with the intent to justify their answer later with the same ambiguity.

Effective RFPs contain process descriptions or user stories that not only list which features are required, but also why they are required and how they are going to be utilized. If you are going to try to select a vendor or system through RFP method, this context will enable vendors to make real proposals and to recognize prospective misalignments prior to contracting.

- Overlooking non-functional requirements

While feature lists are included in the majority of RFPs, non-software requirements generally determine day-to-day user satisfaction and long-term viability. These critical specifications define how the system will impact the operations rather than what it does. Companies routinely omit or poorly define important non-software requirements, such as:

o Performance expectations: Response time, throughput capacity, batch processing windows, and peak load handling are vital to user acceptance and business productivity. Vendors will design systems for average cases unless you provide them with explicit metrics on your volume and performance expectations.

o Security standards: Standards for data encryption, authentication procedures, role-based access controls, and compliance certifications (SOC2, HIPAA, GDPR, etc.) need to be clearly defined rather than presumed.

o Scalability needs: Projected future growth in terms of number of users, transaction volume, data storage requirements, and geographical growth plans greatly increase the likelihood that the selected solution will not be a hindrance to future business expansion.

o Compliance requirements: Industry-specific regulatory requirements need to be spelled out with clear verification procedures since assumptions regarding compliance capabilities are often a source of post-selection surprises. This should never be assumed to be understood by the responder of the RFP. Even if these requirements feel like common knowledge, they must be specifically called out.

o Support needs: Clearly established expectations for support hours, response time, resolution time frames, and escalation procedures avoid misaligned service levels that cause operational disruption.

o Team assigned: Specifically request bios, résumés, or interviews for the team members who will be assigned to your project. Determine if the team will be dedicated or will split time across multiple projects. Ask for commitments during the contracting phase.

Such non-software requirements are a significant gap between potential and real-world success. Leaving them out exposes organizations to the risk of acquiring systems that pass feature checklists successfully, but the implementation performs sub optimally in the real world.

- Premature technical specification

Most organizations, and especially those with a dominant IT department, are guilty of defining technical solutions instead of business needs. RFPs that call for particular technologies, architectures, or methodologies, like "must utilize REST APIs" or "shall be done in .NET," limit vendors from offering potentially superior solutions.

This premature technical orientation sets up a chain of issues. It suppresses innovation as it doesn't allow vendors to propose newer or alternative solutions. It tends to mirror internal IT biases and existing team skillsets rather than real business requirements. Most of all, it deflects interest away from the what and why (business results) and concentrates on the how (technical fulfillment).

The best RFPs are methodologically agnostic but clearly state business needs and desired outcomes. They concentrate on the results, not on how they are to be obtained, and permit vendors to suggest the right technical approaches founded on their experience and capabilities.

- Absent prioritization

When everything is a priority, then nothing is a priority. Yet, most RFPs contain hundreds or thousands of requirements with no relative level of importance. This absence of prioritization causes numerous selection and implementation problems.

At selection time, suppliers struggle to know which capabilities should be emphasized in their proposals and demonstrations. At implementation time, project teams lack a basis for making inevitable trade-off decisions when constraints are uncovered.

Effective RFPs clearly rank requirements with clear prioritization schemes, i.e., must have, should have, could have, and won't have (the MoSCoW method). This kind of prioritization directs vendor responses, informs implementation planning, and offers a basis for scope management in case of issues.

Special Considerations for CRM and ERP Projects

CRM and ERP implementations present unique requirements challenges that are not always present with other types of systems.

ERP implementations have the distinction of demanding a unique set of requirements challenges compared to other types of business software systems. This is mainly due to three fundamental areas: departmental process interdependencies, legacy system constraints, and regulatory or timing requirements.

Process interdependencies are so intricate in the case of ERP due to the fact that a change in a single functional area will automatically affect all departments, causing ripples across the organization. They usually have sophisticated workflows that span departmental lines and depend on master data embedded within them, which need to be meticulously organized and managed. There is also complexity added from legacy systems, since organizations must deal with historic data conversion needs, struggle with the challenging shift away from custom-developed systems to standardized function, and may need to preserve existing integrations to peripheral systems that are not being replaced.

Companies have to navigate a complicated web of compliance and regulatory responsibilities, such as industry-specific demands, geographic differences in conducting business, and strict audit and control demands. ERP implementations, to be successful, require more diligence when building the requirements. It is important to specify:

end-to-end processes instead of standalone departmental activities, process owners and decision rights explicitly, and data conversion and historical reporting requirements exhaustively. They also require care in explicitly stipulating exhaustive testing scenarios for integrated processes as well as compliance and control requirements. In the absence of this method of defining requirements, ERP implementations too often get fragmented and ultimately fail to deliver their intended business value.

CRM projects also face unique challenges when defining requirements, primarily because of three main areas: process variability across individuals or teams, the delicate balance between standardization and needed flexibility, and integration complexity around master data. Process variability might include: different sales teams operating with different processes, geographic or market segment differences resulting in different needs, and individual management styles that are not ready for standardization. Organizations that deploy CRM systems must walk a fine line between standardization (i.e., standardized pipeline reporting and forecasting) and demands for flexibility (i.e., differing stakeholder priorities, such as management demanding data visibility but individuals valuing productivity). Adding to the difficulty is also integration complexity, since CRM systems today are required to integrate seamlessly with marketing automation tools, customer service platforms, quoting and ordering systems, and contract or document management solutions. To transcend these challenges, successful CRM requirements must reflect real-world sales processes instead of theoretical workflows, document the variations between teams or individuals, and pin down possible areas of standardization.

Balance management requirements and reporting and analytics requirements carefully, ensuring coverage of minimum data and data governance prerequisites, and incorporate integration specifications. Companies that do not value these CRM-specific intricacies typically wind up with systems that enforce a stern uniformity and against which users in sales, customer service, and marketing will revolt.

Third-Party Consulting: Balancing the Scales

Many organizations hire third-party firms to develop RFPs, write requirements, and manage the selection processes. This is a common practice with a number of potential benefits but also with significant risks that must be carefully managed.

Requirements documentation and requirements-gathering expertise can dramatically enhance an RFP's quality. System selection firms usually have gone through similar processes with dozens or even hundreds of organizations like yours and can apply this experience to your project.

Market expertise is another benefit since specialist companies possess up-to-date knowledge about software vendor capacity, direction, and overall pricing structures. This expertise helps to establish realistic expectations and prevents the inclusion of ridiculous requirements that no vendor can reasonably meet. They can be a filter for bad ideas.

Most significantly, perhaps, third parties also offer objectivity that is not normally present from in-house groups. Third parties are able to challenge assumptions, pose questions regarding historic practices, and prevent political agendas from corrupting the requirements. Such objectivity also extends to the vendor evaluation stage, where they can apply uniform scoring techniques unaffected by existing relationships.

In spite of such advantages, organizations need to realize that outsourcing RFP creation does not absolve them of the responsibility for the quality of the requirements. The greatest risk is the failure of true knowledge transfer when a third party creates a generic list of theoretical requirements but none of the subtleties appear until there is a close operational analysis.

A similar common trap is template-based approaches, where consultants are applying boilerplate requirements from prior projects instead of taking the time to learn about your specific requirements.

This generates the same problems as internally-developed generic RFPs—lengthy documents that are not specific.

Certain organizations create excessively complex requirements-gathering processes with the intent to prove their organization's worth instead of effectively determining the right solution. These processes can potentially drag out selection, inflate cost, and give the illusion of objectivity by using scoring models, which in reality contain many subjective assumptions.

One sensitive issue is whether companies that assist in RFP preparation should be allowed to bid on the implementation work. Some discourage this to avoid consultants who write requirements to suit their implementation expertise. Others permit it in an effort to achieve continuity from selection to implementation.

There is value in both approaches. Prohibiting implementation bids can increase objectivity at the cost of potentially forfeiting valuable knowledge transfer. Allowing implementation bids helps with continuity but creates possible conflicts of interest that must be managed by open procedures and clear ground rules.

My guidance would be to generally include a clear disclosure of requirements and selection criteria. If firms who develop RFPs are also permitted to bid to implement, use a clear scoring model, have several decision-makers, and demand strong justification of any decision in favor of the RFP developer. Ultimately the departments and stakeholders need to separate the process from the deliverable and to own their requirements rather than relying on outside firms to drive the full selection.

A decision also needs to be made whether to use open or blind bidding processes as an additional strategic choice. In open bidding, suppliers are aware of whom they are bidding against, and this has both positive and negative impacts. Suppliers are able to direct their proposals at the weak points of their competitors, and this could result in better counter proposals. However, it could also lead to intentional undercutting of competitors at the expense of making their own great proposal.

Blind bidding, where the identity of competitors is not known to suppliers, theoretically places greater emphasis on requirements rather than competitive positioning. This method can limit decisions made on the basis of existing supplier relationships. But it can also inhibit suppliers from highlighting competitive differentiators that would be good to understand.

Successful RFP Design: Best Practices

Successful RFPs share several characteristics regardless of the system being implemented.

They are results-oriented instead of feature-oriented, and clearly state success in terms of measurable outcomes. They include the right context information regarding business processes, organizational structure, and strategic goals. They are solution-agnostic but clearly state non-negotiable requirements and priorities.

Above all, successful RFPs set out evaluation criteria that reflect business priorities, containing a common model to contrast vendor approaches that could radically differ. They separate required capability from nice-to-have features, giving vendors an understanding of real priorities as opposed to wish lists.

By dodging the usual RFP pitfalls by adhering to these best practices, organizations that are required or must use RFP processes greatly improve their chances of choosing systems that provide the expected value. This kind of additional effort put into RFP preparation yields dividends when moving to implementation.

Moving Forward

The diligent effort of requirements definition is not a bureaucratic phase; it's the essential foundation that will determine whether your implementation will deliver transformational value or become another data point in the alarmingly high failure rate of technology initiatives. There are no shortcuts to excellence. Finish the requirements work first, then select the technology that best addresses your business requirements.

Clear and well-defined requirements are essential, but they alone do not guarantee a project's success. They provide the foundation upon which successful implementation is built. As you move from requirements to implementation, remember:

- Needs change. Keep in mind the possible need for change to requirements and have a process for it.
- Check early and often with stakeholders to validate the requirements.
- Ensure traceability end-to-end from strategic goals to technical design.
- Promote open, honest discussions on requirements, schedule, and resource compromises.
- Document priorities and their dependencies with the outcome rationales.
- Prioritize business results, not merely technical deliverables or the software itself.

Reflection Questions

1. Have you discovered and incorporated hidden requirements that may be assumed by the users?
2. Are you ranking requirements and managing priority with their perceived and real value?
3. Are you using any methods successfully to elicit requirements within your organization?
4. What is one requirement in your organization that was not written well enough to lead to the right outcomes?
5. Are you looking for ways to enhance cross-department collaboration in the process of requirements and design?

6. Are you identifying and recording the most important non-software requirements for your company?
7. Are you balancing standardization and future flexibility in your system requirements?
8. Are you evaluating both platforms with add-ons and COTS software products?

CHAPTER 8

Don't Forget the Data

"Systemize the predictable so you can humanize the exceptional."
—Isadore Sharp, founder of Four Seasons Hotels

F our million dollars. That's what the organization had invested in the last three years constructing data integrations, and now the consultant was recommending that they remove nearly half of them. The boardroom was silent as executives fidgeted uncomfortably in their seats, passing nervous glances around the room. No one likes to hear their multi-million-dollar investment was wasted.

"These integrations are causing more issues than they're resolving," the consultant went on, gesturing to a data diagram covered in a tangled web of arrows. "Let me illustrate what's going on."

As he reviewed the diagram it uncovered a digital ecosystem in distress. Every night at 3 AM, all systems slowed to a crawl as hundreds of scheduled integrations jockeyed for position. Reports didn't run. Batch jobs timed out. In the morning, IT personnel woke up to a deluge of automated error messages and irate stakeholder emails.

"We're creating digital traffic jams," the Database Administrator explained, "moving so much data back and forth. Nothing gets to where it needs to be on time."

The CIO leaned forward. "But we need these integrations. Our business users require data from multiple systems. We can't have them opening five different systems every morning to see what is going on with their department."

"That's precisely what we need to question," said the consultant. "There is a basic distinction to be made about whether we need data to move. Many of your integrations are in place not because it's the correct solution, but because nobody ever questioned whether there was a better way that does not require the data to move."

This situation plays itself out in organizations worldwide. Firms invest millions in complex data movement strategies among systems, more often than not creating more problems than they are solving. The harsh truth is that many integration issues occur because nobody challenged the assumption that data should move in the first place.

Too often, crucial conversations about data are postponed until the end of an implementation, an item on the priority list that is tacked on once the architecture, features, and workflows are already cemented. This decision can be costly, and it's also consistently the single greatest source of risk we encounter when called to rescue troubled projects. Data is not a technical detail to be determined during implementation; it shapes how information flows, decisions are made, and value is delivered. Planning for data from day one is foundational, and in the instance of a failed implementation, getting back to the starting point to fully reimagine the data structure and integration plan can be the change needed in the project.

The Integration Illusion

Most organizations view integrations through a simplified lens that does not solve the real problems of the organization. The concept seems simple—when something occurs in System A, update System B. Executives like to wave their hands and say, "Just integrate the systems," as if it were as easy as flicking on a light switch.

This simplistic posture views integration as a utility, much like electricity. Customers expect to enter data once and somehow have it everywhere it needs to be. Business users believe developers can simply connect the systems, with little regard for the complexity involved.

The reality is much different. Even seemingly simple integrations, like sending a quote from CRM to ERP, require huge volumes of supporting data to be synchronized: customers, products, price lists, units of measure, shipping terms, discount groups, and more. Every field must be defined, mapped, transformed, and reconciled among all systems involved. One pharma distribution firm undertook what appeared to be a simple customer integration of their ERP and CRM systems. Three months into the project, they realized that customer classification codes, critical for regulatory reporting, were treated very differently in both systems. What was initially envisioned as a two-week integration now necessitated recreating their entire customer master data categorization in both systems. "We just weren't aware that each department defined even simple terms like 'location' so differently," their project manager said. "Sales used ship-to address, finance used bill-to address, and operations used parent company headquarters address. No integration could reconcile those basic differences without redesigning our business processes first."

The Hidden Complexity of Moving Data

The reality of integrations is that they're significantly more complex than most organizations anticipate. Think of all the decisions that must be made for even the most basic integration:

- Which system is the authority for each data element?
- How frequently should data be synchronized?
- Which system wins when the same record is modified in both systems?
- How and who will handle exceptions and errors?
- What validation rules must be applied during synchronization?
- How do you maintain referential integrity across systems so no orphan rows are created?
- What are the existing security and compliance considerations?

Each question opens dozens of others, creating an exponential curve of complexity in most projects if it is not adequately addressed during planning. The result? Integration projects that routinely bust budgets, exceed timelines, and still fail to deliver expected value.

Six Steps to Simplify Your Integration Approach

Before writing a single line of integration code, consider these six steps to ensure your approach is sound:

1. Be honest about the complexity

Resist the urge to over-simplify integrations when communicating with stakeholders. Engage business users in the complexity instead of protecting them from it. Bring them into field mapping exercises, process design, and data ownership decisions.

Example:

> A retail organization embarked on integrating their point-of-sale system and back office ERP system. Developers assured stakeholders it was straightforward and wouldn't require much business involvement. Three months into development, they came to an impasse about returns processing. In the POS system, returns could be processed against the original transaction or as a separate standalone credit with no originating transaction. In the ERP system, returns needed official RMA numbers and approval processes. Neither system could support the other without being rewritten significantly. When the implementation team finally brought accounting, store operations, and customer service organizations together to map the overall returns process, they identified 27 distinct situations that needed to be handled, from price differential exchanges to receiptless returns.

> What had been envisioned as a two-month integration cycle turned into a six-month business process redesign project. By confronting the complexity directly with the respective stakeholders, they were able to develop a solution not only to integrate the systems but also to improve their overall returns process. This positive result would

have been impossible if they had continued to keep the business users oblivious to the technical facts.

2. Reject out-of-the-box thinking

The all too common phrase, "We want the out-of-the-box integration," is a giant red flag. Standard integrations rarely meet specific business requirements without significant customization or configuration. This isn't a shortcut; it's a way of skirting responsibility to properly design your solution. Do a full design and then determine if any standard integrations are a fit for the solution.

Example:

A manufacturing firm selected Microsoft Dynamics 365 Finance and Supply Chain Management to replace its legacy ERP system. During implementation, their procurement director insisted on using the vendor-provided standard Microsoft Dynamics integration with their supplier portal, despite warnings that their unique vendor qualification process may not be supported.

"We're paying for a premium system," he argued. "Surely the standard integration handles something as basic as new vendor onboarding."

Six weeks before go-live, user acceptance testing revealed a critical gap. Their compliance requirements mandated specific vendor certifications to be tracked that weren't part of the standard integration. The firm faced a difficult choice: delay go-live to custom-develop the integration or launch without digital support for their compliance processes and have the staff perform manual data entry in two places for vendor certifications.

The lesson was expensive but clear. Out-of-the-box is a myth that cost them an additional $350,000 in custom development and a three-month delay. Had they approached the integration as a design exercise rather than a checkbox, they could have identified and addressed these requirements early in the project.

3. Start simple and grow complexity

Begin with a minimum viable integration. Move a single record type with just a few fields and the path of fewest exceptions before attempting more comprehensive development. This approach builds confidence, provides early validation, and creates momentum. It becomes easier to account for additional scenarios and add data elements after data is flowing in its simplest form.

Example:

A telecommunications company needed to integrate its new CRM system with its billing platform. Instead of trying to develop comprehensive integration(s) immediately, they phased their approach:

- Phase 1: Basic customer demographics and account numbers synchronized nightly, just enough to allow customer service representatives to look up account numbers
- Phase 2: Added service information and billing status, enabling representatives to answer basic billing questions without switching systems
- Phase 3: Synchronized payment history and statement details, allowing representatives to handle most billing inquiries directly from the CRM
- Phase 4: Implemented real-time service status checking and telephone-based payment processing within CRM

Each phase delivered immediate value while providing insights that will inform subsequent phases. When they discovered inconsistencies in how service packages were defined between systems during Phase 2, they were able to resolve these issues before they affected more complex billing and payment integrations later in the cycle.

By the end of Phase 4, they had a robust, reliable integration that matched their business processes perfectly, something that would have been impossible had they attempted to build the complete solution all at once.

4. Design strategically, implement tactically, and test thoroughly

Take the time to comprehend both source and target systems in depth. Do not assume stakeholders' descriptions of systems are accurate; verify through direct questioning. Document all field mappings, transformation rules, and error handling procedures in writing. Deploy based on your documented design, not based on developer preference or workarounds implemented to address specific testing scenarios. And above all, test thoroughly, including edge cases and error conditions that will be faced after go-live.

Example:

A financial services organization was integrating its loan origination system with its core retail banking system. Their implementation team meticulously documented field mappings and transformation rules against business requirements, creating a comprehensive design document.

At implementation time, developers realized that there was a performance bottleneck in handling very large bundles of loans. Instead of following the agreed-upon design, they introduced a workaround in which data was batched in a different way, satisfying the requirements functionally but violating the data integrity model.

When the integration was launched, everything appeared to be working correctly until month-end reconciliation, when accounting noticed that some of the loan data wasn't being synced. The ensuing investigation revealed the implementation variance, which had introduced timing problems that occasionally caused batch data to fail.

The company had to go dark with their integration for a month and re-implement along the lines of the original design. This strengthened their policies to adhere strictly to the design-implement-test cycle, with the addition of a formal code review process to keep decisions in line with design even when testing suggests the need for quick workarounds.

5. Establish definitive systems of record

Where possible, for every data table, designate one master system source. Avoid the urge to turn on two-way updates for the same data, which will lead to cascading synchronization conflicts. This becomes challenging for organizations, not because of data, but because it will require the implementation of new processes that may not have existed before. Accounting used to be able to update customers, and the sales team could update customers. Now we will only allow one group to 'own' the customer data and implement processes for how we handle updating future records.

Example:

A retail chain tried to synchronize product information on its web storefront and in-store point-of-sale system. They started by enabling product changes in both systems, figuring the latest change would win.

This approach quickly created confusion. The retail managers would change price lists in the POS based on local competitors and in-store promotions, while the e-commerce team would change the same products based on online competitors and e-promotions. Product descriptions were also changed independently in each system.

After months of reconciliation nightmares, they redesigned their approach. The ERP system was established as the authoritative source for product information, with one-way synchronization to e-commerce and POS. Store manager and e-commerce group change requests were routed through a centralized product information change request process.

This final data ownership arrangement took away the conflicts, brought consistency to systems, and even improved their ability to respond to market situations because decision-making was centralized and deliberate rather than fragmented and reactive.

6. Implement the correct pattern

One size does not fit all. Different goals for integrations demand different technical solutions. Your pattern of implementation must fit your business needs instead of a general solution. An integration solution designed based on the particular skills of an in-house developer rather than based on the right solution to the problem is a particularly bad outcome.

Example:

A shipping organization required its drivers to get delivery details from its transportation management system while on the road. Their first implementation was a data polling design pattern; the mobile application polled for data changes every five minutes and transferred the changed data into TMS.

This method had two issues: drivers occasionally missed crucial delivery updates (if updates happened between instances of polling), and the polling frequency depleted device batteries and consumed mobile data quickly.

Once they examined the underlying business need, providing drivers with real-time delivery information, the firm moved to an event-driven model based on push notifications. The TMS pushed instant updates for critical events and less frequent updates for non-critical events, but only when delivery information had changed, providing drivers with constantly updated information and preserving device battery and data usage.

By selecting an integration pattern that most aptly met their specific business need rather than relying on the familiar polling method, they improved technical performance and business outcomes. The key was focusing on what they were trying to achieve rather than starting with a technical solution simply because it is the way it had always been done.

Identifying the Real Requirement

The key to a successful integration strategy is to look beyond the initial request to fully identify the underlying business requirement. When someone says, "We need an integration," the probing question is, "What problem are we actually trying to solve?"

The answer typically falls into one of these categories:

- Single point of data entry
  - *Example*: "I just want to enter customer details once."
  - This is a valid integration reason. You're avoiding duplicate data entry and providing consistency. If users input data into one system and it's converted in some form into another, data movement is generally required.
- Process integration
  - *Example*: "Once a lead is qualified in marketing, I'd like it to be automatically moved into the sales pipeline."
  - These event-based integrations facilitate a specific business process that extends across several systems. They're usually event-specific rather than synchronized continuously. When this occurs, if the criteria are true, then move or create a subset of data in a different system.
- Data synchronization
  - *Example*: "I want my product catalog to be the same in both systems."
  - This requirement is about consistency of reference data between systems. It is generally required for subsequent transaction processing. Generally, there is no user

intervention needed. The data just needs to be moved, mirrored in multiple systems.

- Single pane of glass or dashboards
  - *Example*: "I want to see all customer information in one place."
  - This is where companies typically make significant mistakes. This need doesn't always call for data movement when there is a need for high-visibility data. There are many ways of exposing data and unified views without copying data from one system to another. Creating read-only data views or consolidation dashboards can eliminate data synchronization issues and simplify the integration processes.
- Legacy system access
  - *Example*: "We need to keep our historical data and reports available to users."
  - This is seldom a good reason for integration. There are other ways of ensuring access to old data without introducing it into your new production environments. Implementing a data lake, archiving historical reports, or building constant tables where data can be viewed but never changed can provide visibility without the complexity.

Understanding what type of integration your requirement falls under allows you to select the appropriate strategy. Does the requirement require traditional data movement integrations or can they be addressed in other ways?

Beyond Data Movement: Alternative Integration Patterns

If the business need is mostly to have visibility to the data but not to transact against it, standard integration approaches that move data are excessive. Data movement integrations bring in a high degree of complexity, volatility, maintenance overhead, and potential performance problems that should be avoided.

Prior to reflexively constructing a data movement integration, pause to consider some important questions. Is this actually a transactional requirement, or is this simply a case of exposing data? Would slightly stale or older data still satisfy the business requirement in most cases? How frequently will this data actually be changed and utilized, and does that justify syncing it on a periodic basis? What are the data volume implications of moving this information, and could a report satisfy the requirement with far less complexity? If your analysis shows data visibility to be the main need, consider alternative methods to simplify your patterns.

Reporting Solutions

Business intelligence technologies such as Power BI or Click Report can simultaneously connect to multiple data sources and provide unified views without transferring data between operational systems.

This strategy offers the following pros:	This strategy offers the following cons:
No duplicated data storage, no synchronization problems created by timing	Limited to no write-back capability
Real-time access to data directly from source systems	Performance can degrade with very large data volumes
Robust data visualization features that emphasize cross-system relationships	May involve extra licensing or software expenses
Less development and maintenance costs compared to conventional data integrations	May require UI/UX skill sets that are outside normal development teams
	Less suited for operational processes that require data manipulation

A distribution company that was experiencing difficulties with the visibility of inventory data between their ERP and warehouse management systems had initially conceived of a bidirectional integration. This proved to be fraught with complexity. They ended up implementing a Power BI solution that pulled real-time data from both systems, providing warehouse managers with rich dashboards of on-hand inventory, open orders, and inbound shipments, without actually moving a single record between systems.

Apps

Developed applications using low-code solutions can be used to craft custom interfaces that bring together data from various sources into a single view. They can also enable updates back to source systems.

Pros	Cons
Highly customized user experiences created for targeted business requirements	Can present issues of application governance if not addressed early
Can mix read and write ability as needed to solve business problems	May become too complex to support as requirements change
Can often be developed by business power users with little IT intervention using low code tools	Performance issues can occur with large amounts of data
Support for workflows or business processes that span multiple systems	Possible security issues if not well designed with security in mind
	Risk of spawning new shadow IT departments if not governed appropriately

A services company created a Microsoft Power App Canvas application that showed customer data from their CRM, service history from their custom field service application, and billing data from their SAP ERP. The service representatives could see this consolidated data and even process returns or credits from within the app, which would write back to the underlying system behind the scenes.

Embedded Web Experiences

Modern business applications support embedding external web applications within their interface. This allows you to present data from other systems in context, creating a unified experience without integration complexity.

Pros	Cons
Minimal development effort in comparison to conventional integration	Limited integration with host application (mostly support for data visibility only)
Leverages existing web interfaces Embedded application updates are displayed in line	Authentication and security issues between apps and domains
Suitable for rarely accessed or historical data May offer rich interactive experiences spanning multiple systems	Inconsistencies in the look and feel of application user experience
	Dependency on external application availability and reliability
	Possible browser security restrictions that cause issues

The correct strategy to choose will be based on your particular requirements, but for most situations, these alternatives can provide the necessary business value earlier, at less expense, and with less risk than conventional data movement integration, which can add risk and issues to an implementation.

The Right Integration Pattern for Your Needs

When data movement is unavoidable, the right pattern for moving data through integration needs to be selected. The various patterns each possesses unique benefits and constraints that render it fit for specific circumstances. Below are a few common patterns to consider based on the volume, frequency, and criticality of the data and systems being integrated.

Point-to-Point Integration

A point-to-point integration provides a one-to-one connection between two systems where one system calls APIs or the service layer of another system directly. The integration pattern is specifically designed to the respective systems being integrated, with no intermediary layer for handling the communications. This method is appropriate for straightforward, direct integration between two systems, especially when there are real-time updates required for business-critical processes with low to moderate transaction volumes. Implementation is typically quicker with minimal infrastructure requirements, making it a good fit for event-driven scenarios that require immediate response, i.e., creating a customer in ERP when the quote in a CRM has moved to a certain stage. However, it's a troublesome approach when dealing with multiple systems as it leads to a difficult "spaghetti" infrastructure that's tough to support. It's also not suitable in cases with complex routing or data transformation, when you anticipate changes in endpoints, or in extremely high-volume data transfer situations that could impact performance.

Service Bus Pattern

The service bus pattern, typically a middleware architecture pattern, contains a message broker in the middle that handles communication among different systems. The messages are routed to the service handling systems and then distributed to their destinations according to subscription patterns or content-based rules defined by the subscriber.

It is optimally suited for integrating more than two systems for complicated routing situations where content-based or rule-based delivery of data is necessary. It offers assured message delivery with a retry pattern for failures and can be suitable for large data volumes because the bus can queue messages.

This pattern supports transaction capability across many systems, and can be perfect for scenarios such as routing orders to various fulfillment systems based on product, customer location, or warehouse availability. This pattern is less suitable for two-system integrations where the overhead of the bus does not make sense, when the lowest latency possible is needed (since the message broker introduces processing time), for small companies with minimal IT infrastructure resources dedicated to maintenance, or when real-time synchronous responses are required.

Polling Pattern

Polling pattern(s) utilize an approach in which the target system checks the source system for new information or changes at scheduled intervals, usually on a scheduled job running at a fixed interval. Polling is great for data that does not require real-time updates, when source systems lack the capability of event notification or webhooks, or where bulk synchronization is done for the purpose of building reference data, such as product and prices. It's especially helpful when dealing with legacy systems that don't have modern API functionality or integrating with third-party systems where the source can't be changed (i.e., nightly syncing of customer master data between systems).

Avoid this pattern when business-critical updates must be in near- or real-time, for high-frequency data change that would require high-frequency polling cycles, in high-volume scenarios where polling causes a backlog of performance issues, when the source system changes by API call (so polling is expensive), or when immediate response to events is required for business processes.

Broadcast Pattern

Integrations that use the broadcast pattern are based on a publish-subscribe model where one event in one system notifies multiple recipient systems all at once. The source system broadcasts updates to any interested subscribers without having to specifically know who they are. This pattern is applicable when one change must be published to various systems, for ensuring consistency of data in distributed systems, or when the recipients have different processing needs of the same data. It is an adaptive pattern where new subscribers can be added without changing the source and it promotes loose coupling of systems, which makes it ideal for requirements like integrating the available quantities of products whenever the stock levels are updated. This pattern can be troublesome when you need to guarantee processing by all recipients before proceeding with a transaction, when there are complex transformation requirements specific to each recipient, when strict order of operations is needed, or when working with sensitive data that should not be broadcast to a centralized system.

There are many right answers, but your pattern must fit the particular business need, technical limitations, and organizational capabilities. The most effective integration approaches may use several different patterns. The art lies in applying the correct pattern to each individual integration instead of imposing a single approach to everything. This is where so many organizations find themselves in a rescue situation due to a spiderweb of integrations moving data in various directions at various times.

Case Study

Richard, the incoming CFO of a major global home goods distributor, opened the board meeting he requested. "Good morning, all. I'd like to talk about something that's been hindering our company in a big way—our visibility into data. We're currently making multi-million-dollar decisions based on spreadsheets, gut instinct, and contradictory reports from various departments. This is not sustainable.

"At my former company, we revolutionized our business with SAP S/4HANA. We moved from disjointed, inconsistent, manual reporting to real-time analysis, automated reporting, and real predictive forecasting. We could see the sales forecasts, inventory availability, and production costs at a glance.

"Here, we just don't have this functionality. Our ERP system is not giving us the information. We're in the dark.

"My recommendation is that we do a total rebuild. SAP S/4HANA will modernize our reporting, consolidate our supply chain data, and provide leadership with real-time dashboards and analytics. It's an investment, I know, but it is one that will position us for long-term success."

The boardroom was quiet after the new CFO concluded his presentation. Slides showing dashboards with analytics, color-coded charts, and valuable reporting remained on the screen, a digital utopia that was the opposite of their current reality that consisted of fragmented Excel reports and manual data pulls.

"This is visibility," stated the CFO, gesturing towards the screen. "This is what SAP provided for my last company, and this is what we need here."

The CFO had been at the company for just six months, having been headhunted by them from his previous position where he'd run an award-winning digital transformation program. Ever since he joined, he'd been grumbling more and more about their stone-age reporting.

"Our current ERP system is holding back the company," he continued. "I'm flying blind. Sales projections, inventory valuations, production costs—I'm getting different numbers from every department. We're making decisions off of bad data and hunches."

The CEO, Elizabeth, nodded thoughtfully. She'd lived it herself, waiting for days to get the answers to what appeared to be straightforward questions about business performance, and then getting a different answer from another department.

"This is what you suggest, Richard?" she inquired.

"Yes!" answered Richard, "Total replacement. The SAP S/4HANA platform revolutionized our visibility into operations. We'll have real-time dashboards, automated analysis, predictive forecasting, and everything we lack today."

Within two weeks, an RFP was issued and a search for an implementation partner had begun. Three months later, a selection committee recommended proceeding with the SAP implementation at an estimated cost of $13.5 million, plus internal resources and cost business disruption. The steering committee scheduled a final review meeting before presenting to the board for final budget approval.

The night before this pivotal meeting, Elizabeth sat in her office, staring at the implementation timeline. Something didn't feel right. The company had invested heavily in their current ERP system just three years earlier. While not perfect, it had certainly modernized their operations at the time. Could it really be so inadequate now?

On impulse, she called an independent consultant she'd worked with years earlier. "I need a second opinion before we commit to a $13.5 million system replacement," she explained. "Can you take a quick look?"

He arrived the following morning, sitting quietly through the steering committee's final review. The implementation timeline stretched across 36 months. The risk assessment highlighted potential disruptions to every department. The change management plan included significant retraining for 400+ employees. The ROI analysis promised savings that would recoup the investment within four years.

When the presentation concluded, he asked a question that would change everything: "What specific data are you unable to access today?"

The room fell uncomfortably silent.

"All of it," Richard finally responded, gesturing toward his sample dashboards. "Real-time profit analysis by product line. Integrated supply chain metrics. Labor productivity by work center."

"Those sound like measures and outputs," the consultant noted. "I'm asking about the underlying data. Is the information actually missing from your systems, or is it just difficult to access and visualize?"

Over the next two weeks, the consultant led a focused investigation, working closely with the IT team and business analysts. They examined database structures, integration points, and data flows across the organization. They interviewed department heads about their information needs, why various Excel systems were being used, and the manual reporting processes.

The findings were revealing:

- Nearly every data element the CFO wanted was already captured somewhere in their systems. The inventory management module tracked stock levels in real-time. The production system recorded labor hours and material consumption for each job. The sales platform maintained detailed pricing and margin information.
- The problem wasn't missing data; it was fragmented, inconsistent, and inaccessible data.

"Your issue isn't your ERP platform," the consultant explained to the executive team. "It's your data governance and reporting infrastructure."

He outlined the real problems:

- The same business terms were defined differently across departments. "On-time delivery" meant something different to production, logistics, and sales teams.
- Master data was maintained inconsistently. Product codes didn't match among systems, requiring manual translation in the various Excel reports.
- No single repository combined data from all operational systems. Reports required manual extracts and reconciliation every time they were generated.
- Business users lacked self-service analytics tools, relying instead on IT-generated reports that took days to produce.

"What you're actually lacking is the analytics platform and data warehouse that sat on top of your former SAP environment," the consultant clarified to Richard. "That environment wasn't vanilla SAP; it was a custom design developed over the course of years following the core SAP implementation."

Richard appeared amazed. "You're telling us we don't need to replace our ERP?"

"Not for data visibility and reporting," the consultant concurred. "You need a good data plan, some modern analytics software, and maybe a data warehouse."

Six months later they went live with their new business intelligence solution. They had implemented a data warehouse that provided departmental-standard definitions, rigorous master data management, and Power BI dashboards that provided almost the same features as the CFO's vision for SAP.

The price? $875,000—less than 15% of the proposed ERP replacement.

Most of all, they experienced benefits quickly instead of 36 months down the line. Day-to-day operational dashboards for executives were up and running within a few months. Department managers themselves could dissect performance trends without IT involvement. Sales forces gained production capacity and inventory visibility in order to manage customer expectations within a sales dashboard.

"I was focused on the system, not the data," Richard conceded six months later, scrolling through the same metrics he'd displayed in his original presentation, now filled in with their live data. "At my last company, I wasn't looking at the bare SAP system. I was looking at years of data refinement and analytic development."

Elizabeth, the CEO, smiled. "And if we had replaced our ERP? We would have invested millions to take a step backward."

Richard conceded. "Despite flawless execution, we would have most likely lost our historical data comparisons and spent years recreating the reporting maturity we have now."

This case is a valuable lesson about enterprise systems. They had a good ERP in place at the time, but it excelled at only continuously capturing transactional data that ran the business. They needed a more effective data consumption layer, governance and mechanisms to translate data into actionable intelligence. By concentrating on the outcome (improved business visibility) instead of the supposed solution (ERP replacement), they attained their objectives more quickly, less expensively, and with much less disruption.

Moving Forward

The future of data and integration is not quicker movement with more data; it is smarter use of less data. As companies embrace these architectures, API-first design patterns, and event-driven systems, the old notion of integration must change.

Modern integration approaches focus on:

- Communicating that something happened rather than copying data and guessing what may have happened
- Interacting with systems through published interfaces, apps, artificial intelligence models, or analytics platforms rather than accessing their data directly in a system
- Building modular capabilities that can be used to handle a specific organizational use case, rather than developing and implementing major systems
- Focusing on end-to-end business processes that may span multiple technologies

The above reflects a radical shift in the way we need to approach linking systems, from moving data to collaborating on services. The most effective organizations will be those that grasp these concepts, with less emphasis on replicating data among systems and more on coordinating processes across them.

Reflection Questions

1. How many integrations do you have in your environment today that exist simply to provide visibility to information?
2. What business processes cross more than one system at your company, and how do the handoffs get coordinated today?
3. For each of your existing integrations, can you articulate the definitive business requirement it addresses?
4. What alternative approaches could you consider for providing users with cross-system visibility without moving data?
5. How do you currently manage master data across systems?

---

6. What governance processes exist for approving and implementing new integrations in your organization?
7. Do you measure the success and business value of your integration investments?

CHAPTER 9

Training, Upskilling, & Reskilling Your Team

"The only thing worse than training your employees and having them leave is not training them and having them stay." —Henry Ford

"Our ERP vendor has just let us know that our lead consultant is leaving the company. They're putting in a new resource next week." The announcement came from the CIO.

Silence.

The project sponsor took a deep breath. "Mike has been with us for eight months. He knows our business inside and out. This will set us back."

The CIO wasn't finished. "And those custom reports we've been waiting for? The vendor now tells us they'll be triple the cost because they're more complicated than they expected."

The CFO leaned in, fingers drumming on the table. "We're months behind schedule and 30% over budget already. How much more are we going to pour into this?"

This wasn't just a bad day. It was an all-too-common nightmare. Companies invest millions in enterprise systems, only to be left stranded, reliant on third-party consultants who, as the only ones who understand the tech, hold the keys to it all. When those consultants leave, they take with them essential knowledge, and the company scrambles.

This may sound ironic coming from me, one of the managing partners and founder of a consulting firm that specializes in enterprise system implementations, but the reality is that the best consultants don't want to be stuck in your company forever. The best consultant personalities want to deliver value, transfer knowledge, and move on to other challenges. The consultant personality type is most frequently motivated by variety, a chance to solve different problems, deal with different technologies, and experience different business environments. For a consultant, becoming the person responsible for the day-to-day maintenance and management of one client's system so it will be running forever isn't only bad for the client, it quickly becomes boring for the consultant.

The issue is not consulting in itself. The issue is most organizations lack the capability of building internal competency. Businesses that manage to develop in-house know-how not only reduce costs and risks, they transform their enterprise systems from expensive technical burdens into a competitive advantage. They make their technology a strategic asset.

In earlier chapters, we covered the selection of the right systems, stakeholder alignment, and project team structuring. However, even the finest implementation is doomed to fail if your organization is still dependent on external experts for deep understanding of your environment, daily operations, maintenance, and growth. This chapter is your roadmap to breaking that vicious cycle. We'll discuss how to build in-house expertise, create star performers, craft training programs that work, and build a culture of continuous learning. At the end of the day, true success isn't about launching systems; it's about taking ownership of them.

This critical aspect of digital transformation is where many companies go wrong. Just as 70% of digital transformations fail, only 16% of companies report that they've truly improved performance and set up their organization for long-term change (McKinsey, 2018). What's the missing link? Investing in the people to power the systems, not just the technology.

Too many companies want to upgrade their business applications, but they neglect upgrading their people. This oversight creates a vicious circle: as technology evolves, the knowledge gap widens, leading to frustration, disengagement, and eventual attrition. Companies are then forced to bring in new individuals at top-market salaries, individuals who have developed exactly those skills somewhere else rather than building the same competencies internally.

The numbers tell the story. PricewaterhouseCoopers research shows that employees who don't get enough upskilling opportunities are almost three times more likely to seek new employment compared to those at firms with robust upskilling programs (PwC, 2024). An equal investment in systems and people is needed for successful digital transformation. When organizations fail to invest in either aspect, they lose the long-term value of their technology investments and risk building an expensive constant recruitment cycle.

Why Internal Ownership Is Critical for Long-Term Success

Why is this mindset critical? Enterprise systems are not static but adaptive systems that need to evolve continuously. Business needs don't stand still, market pressures evolve, competitive forces intensify, customer needs change, and regulatory needs shift. What is sufficient today will be insufficient tomorrow. Also, technology and the platforms themselves march forward, with new functionality, security patches, and design enhancements that organizations need to assess and possibly introduce. The business itself is also constantly evolving, acquisitions and mergers, new lines of business, and geographical expansion will all require system changes. A CRM system today will be supporting a substantially different enterprise three years down the line. Firms that do not implement people and skills before long will discover their investments paying decreasing dividends, with workarounds in abundance, adoption behind, and competitive advantage dwindling. Within such a continuous transformation culture, the ability to evolve your enterprise systems is as important as the initial configuration itself.

This inability to evolve shows up in a number of ways:

- Reactive vs. proactive improvement: Without internal expertise, organizations are forced to react to issues instead of proactively improving their systems. By the time they bring in an outside consultant, the business opportunity may have passed.
- Knowledge fragmentation: Key system knowledge is dispersed among different consultants, suppliers, and documentation methods. Nobody has an end-to-end understanding of how the overall system facilitates business operations.
- Increasing costs: Reliance on outside expertise turns into a cost driver as routine maintenance tasks, minor improvements, and even simple reporting necessitate costly consultant time.
- Implementation amnesia: As time passes, the reasoning for the most significant configuration choices is forgotten and organizations are left with a lack of knowledge of their own implementation and systems.
- Decreased business alignment: Systems gradually stop aligning with business needs because the cost and complexity of using third-party consultants discourage small, evolutionary modifications that could be handled internally.

This scenario played out in a manufacturing company that implemented a comprehensive ERP system:

"We spent $4.5 million implementing our ERP system, but three years later, we were still spending over $800,000 annually on consulting hours," explained their IT Director. "Every time we needed a new report or workflow change, we had to engage consultants who charged premium rates. We had no alternative.

Even worse, they took weeks to respond because we were competing with their other implementation projects for resources."

The company decided to establish an internal ERP center of excellence, investing $350,000 in training key employees across IT and business units. Within one year, they reduced external consultant spending by 70% while simultaneously increasing the pace of system enhancements by 40%.

"The ROI wasn't just financial," the director continued. "When our production scheduling needs to be changed due to a major new customer, we are able to modify our system in weeks rather than months. That agility gives us a competitive edge that would have been impossible with our previous consultant-dependent model."

Internal ownership does not mean the complete exclusion of outside assistance. It means altering the relationship from dependency to collaboration. This mentality shift fundamentally changes the power balance between your company and external consultants. Rather than negotiating from a position of weakness where consultants have exclusive knowledge of your operating systems, you negotiate from a position of information. Your personnel understand the system architecture, configuration, and integration points, and as a result, they are able to evaluate consultant proposals objectively and so you can make your own choices. You also have a better understanding of the actual effort required when evaluating the costs and benefits of an enhancement.

Within this model, consultants are used as accelerators and specialists instead of gatekeepers. They may still be required for upgrades where extra resources are needed, for development where technical expertise is needed that your organization has yet to gain, and for strategic advice that extends your viewpoint. Most importantly, your company controls when and how these interactions take place, initiating them when they fit instead of when system problems require an immediate response. You control the level of consultant involvement, precisely defining the engagement scope instead of leaving it open-ended. And you control the nature of the relationship, establishing strict knowledge transfer requirements and documentation standards so your internal skills can increase with each interaction. The outcome is a healthier, more effective partnership whereby consultants augment your in-house capabilities rather than owning them.

This transition from dependency to ownership requires three key elements:

1. Establishing a center of excellence that serves as the hub for system education, governance, and continuous improvement
2. Identifying and developing internal experts through targeted upskilling and reskilling programs
3. Creating a culture of continuous learning that expands over time

Let's explore each of these elements in detail.

Establishing a Center of Excellence (COE) for Business Applications

A center of excellence is essentially a defined team of system experts that transform your systems into a business problem-solving engine. It is not a support desk or just another support organization, it is a concentrated team that knows your business processes as much as your technology environment. The people on this team must also have a strong understanding of the change processes within the applications they own. Within a COE, team members have the skills to interpret what the business requires, what is possible with the technology, and the governance for deliverables both functional and technical.

The initial step in creating an effective COE is the clear determination of its purpose and form. This will guarantee that the COE provides value and does not turn into yet another set of meetings or an administrative burden.

A successful COE accomplishes these important things for your organization. Here's how to implement each one:

- Process governance: Establish well-documented, clear standards for how your systems are to be used and changed. Do this by instituting a review team consisting of representatives from each department that meet on a regular basis to weigh change requests versus business priorities. Establish approval flows with criteria that need to be satisfied prior to changes being made.
  - *Example*: A manufacturing firm set up a monthly change control board with representatives from production, sales, finance, and IT. All requests for system changes required a documented business case, risk analysis, and rollback plan. When a production manager asked for a change in the inventory process, the board considered how this would impact financial reporting and sales order processing before giving its approval, to avoid a single solution with a local fix that would create issues downstream. Each change followed a rigorous design and documentation process that was reviewed by the team and approved periodically as changes were requested and the cycle of development and implementation continued.
- Knowledge management: Create a central repository or document management system where system documentation (i.e., L1–L6 documentation discussed in previous chapters), training guides, and process manuals are kept and updated on a regular basis. Require a standard template that all team members follow when they request or implement systems or solutions. Hold knowledge-sharing sessions where internal team members give presentations on particular system features.
  - *Example*: A non-profit organization created a SharePoint site with a separate section for every module of their ERP system. When they revamped the grant management process, they substituted the documentation right away with step-by-step guides and screen shots. Six months later, when the key accounting person quit, her replacement was able to learn the tailored process using these resources rather than bringing in a third-party to do costly training.

- User education and support: Establish tiered support levels with defined escalation procedures. Set response time expectations (i.e, 4 hours for critical problems, 24 hours for standard requests). Utilize a tracking system where users can log issues and see updates. Monthly analysis of support requests that have been logged will create the next month's training requirements and system upgrade opportunities.
  - *Example*: A retailer instituted a three-level support model for their POS software. Associates contacted their "super user" peer first for everyday problems. If unresolved, requests moved to the internal IT support via a ticketing system and triaged the correct members of the COE. Only genuinely difficult technical problems made it to tier 3, the third-party vendor support group. This model resolved 70% of problems at tier 1, lowering support costs by 45% while trimming average resolution time from days to hours.
- System enhancements: Create a cadence for system improvements, monthly for incremental improvements, quarterly for larger changes. Implement a user feedback system (surveys, focus groups, or in-app feedback forms) to discover pain points. Develop a prioritization framework that considers technical complexity, business impact, and resource requirements (L1-L4 requirements referenced in previous chapters).
  - *Example*: A distribution business employed quarterly release cycles for their WMS improvements. Ideas from business users were collected via a SharePoint list that made them log the anticipated benefit (time saved, errors minimized, etc.). All requests were scored by the COE team on a weighted matrix based on strategic alignment, user impact, and effort of implementation. This took the place of their former method where the loudest voice or highest senior requester most often received the highest priority.

- Training and development: Use a mix of training delivery methods (face-to-face, remote, self-paced) to suit different team member learning styles. Develop an in-house certification program that acknowledges proficiency levels. Provide refresher training at regular intervals to fill skill gaps detected through support requests or as new features are released.
  - *Example*: An organization that offered financial consulting and auditing established three certification levels for its CRM system. The "essentials" level addressed basic navigation and everyday tasks for all users. "Advanced" training addressed reporting and dashboards for managers. "Expert" certification addressed basic system configuration and workflow administration. Completion of each level unlocked further system permissions and was linked to performance evaluations. Internal support requests to IT fell by 60% and those proficient in advanced features grew by 40% in a single year.
- Vendor management: Assign individual team members within the COE as owners of vendor relationships. Hold quarterly business reviews with vendors to cover roadmaps and problems. Develop a vendor performance scorecard to measure responsiveness and quality. Have clear contractual knowledge and document expectations for all vendor engagements.
  - *Example*: A company in the professional services sector appointed its top IT manager as its Microsoft Dynamics relationship owner. They instituted quarterly business reviews with their Microsoft partner, whereby they tracked against defined metrics: response time, quality, and knowledge transfer. When the partner consistently missed SLAs, the relationship owner secured approval to get a dedicated resource for the next quarter, turning a deteriorating relationship into a successful partnership.
- Innovation facilitation: Establish an innovation lab in which users can test new features and capabilities in a sandbox

environment. Hold quarterly innovation days where teams present innovative solutions to business challenges or new features they have learned about in the last quarter. Establish a formal sponsorship process and define how innovations go from concept to production.

- o *Example*: A chocolate candy manufacturer and distributor implemented an innovation lab in which COE team members were allowed to spend up to four hours per week learning new features or experimenting with new solutions to ongoing business issues. A department spent the time building a Microsoft Power App™ that automated the escalation of feedback that needed a response on social media platforms. What began as one experiment grew into an enterprise-wide solution that increased customer satisfaction annually.

A centralized COE brings all system expertise and governance into a single corporate team that serves the whole company. This model provides consistency and efficiency, but can put expertise at a distance from daily business processes. A decentralized model disperses COE duties among business units, putting expertise near business users, but can lead to inconsistencies and duplication of efforts. Most companies discover that a hybrid model, with centralized standards and governance along with dispersed expertise, offers the best of both worlds.

Let's discuss the benefits and challenges of each model in detail:

Centralized COE

In this model, a single team is responsible for supporting the enterprise system across the entire organization, maintaining control of everything related to a set of applications or workstreams.

Advantages:

- Consistent standards and processes

- Efficient resource utilization
- Clear accountability and governance
- Comprehensive view of the systems they support

Limitations:

- Potential distance from business unit needs as the organization grows
- Possible struggles with specific requirements in a geography or department
- Risk of becoming a bottleneck if growth overwhelms the skills or team assigned

For instance, if a multinational financial services company has a centralized COE for its CRM implementation, then it can be certain that all users adhere to the same set of procedures in handling customer information, regardless of geographical location, which is essential to ensure regulatory compliance across nations. However, they can also lack deep understanding of regulations that exist in certain countries or departments that may have specialized processes or procedures to address exceptions for their area.

Decentralized COE

This model distributes COE responsibilities across regions, business units, or departments, with each developing expertise relevant to their specific needs.

Advantages:

- Deep understanding of requirements specific to business units
- Faster response to local needs and generally smaller teams
- Greater business unit buy-in and personal relationships
- Specialized expertise for unit-specific processes within a system

Limitations:

- Potential for inconsistent standards and processes

- Redundant resources and effort where different teams solve for the same general problem
- Fragmented knowledge across departmental or business units
- Potential for disjointed evolution where some groups innovate faster than others

For instance, if a healthcare system with separate hospitals, clinics, and specialty centers decentralizes their billing system COE, then each of the facilities can retain specialized competency in line with their individual clinical workflows while they organize to uphold enterprise-wide standards. However, there is a risk that some hospitals may move slower than some clinics and create innovation disparities and technical debt for the organization as a whole.

Hybrid COE

The hybrid approach takes some aspects of each, usually with a central COE setting standards and control, and business units retaining specialist expertise, design, and development.

Advantages:

- Balance between global consistency and local flexibility
- Possibility of hyper specific resource allocation
- Detailed expert knowledge at both organization and business level
- Scalable solutions for different sizes of organizations

Limitations:

- More complex to manage governance separately
- Generally more team members and role redundancy
- Must have definite role specifications and decision-making hierarchy
- Added effort for coordination across COEs
- Greater communication overhead and need for program management

For instance, if a manufacturing company with multiple divisions uses a hybrid strategy, the worldwide COE may define companywide data standards, release cycles, document templates, and security protocols while divisional specialists concentrate on procedures, reporting, and workflow needs particular to their product lines.

For most organizations, the hybrid model is the optimal compromise between standardization and specialization. It offers the advantages of both centralized and decentralized structures without their disadvantages. Hybrid structures are most valuable for multi-entity firms where business units, companies, or departments possess unique operating requirements but still need enterprise-wide consistency.

The hybrid model succeeds because it acknowledges a fundamental paradox in enterprise systems management. Systems need to be both technically sound and business-relevant. Technical standards, security protocols, and data governance need to be centralized to limit risk. But the implementation of specific business processes needs intimate knowledge that can only reside in business units. The hybrid model reconciles these two elements by keeping under central control (i.e., reporting to a technology department) what needs to be consistent and decentralizing what needs to be specialized. This works well as long as the team is cohesive and has aligned goals. In highly political organizations where there is much work to be done around leadership alignment, a centralized or decentralized method could be the most effective.

The hybrid method is particularly useful for organizations that have:

- Several business units with varying business models
- Different geographic areas with different regulatory demands
- A blend of older and newer systems that call for different skills
- Growth through acquisitions or mergers leading to disparate system landscapes

As an example, a global manufacturer can have a central COE team working on system architecture, security, and overall configurations, but with specialized COE members embedded in each business unit. Whereas the central group is tasked with all instances of following the same set of upgrade timelines and security protocols, the embedded specialists possess department configurations specific to each manufacturing workflow in each division. The key to effective hybrid models is to establish well-defined roles, strong leadership, clearly stated responsibilities, and solid coordination so there are no gaps or overlaps. Explicit separation of central versus distributed responsibilities must be documented and revisited periodically, and there should be formal communication channels between central and embedded teams to achieve deep alignment.

Core Disciplines: Governance, Training, Support, and Innovation

No matter the structure, every COE needs to develop some essential disciplines within the organization. We have used the word disciplines here instead of responsibilities because these are not some functional duties to be checked off; they will need to become ingrained in the organizational culture in order to succeed.

These fundamental disciplines have a life outside the COE itself and need to establish the way that everyone dealing with enterprise business systems will work. If restricted, they will be bottlenecks rather than enablers. If the disciplines are instilled across the organization as cultural norms, they will ensure a culture of continuous improvement.

Let's talk about each of these disciplines and how they must become ingrained in your organizational DNA.

Governance

The COE establishes and enforces the rules, standards, and processes that govern how the enterprise system is used and evolves.

Critical governance elements include:

- Data quality standards, audit, and monitoring

- Change management processes
- Configuration management processes and release schedules
- Security and access control for all users
- Integration standards and methods
- Development and testing patterns
- Release management and ALM processes
- Management of the chain of change approvals

A good governance model keeps the system from descending into anarchy while accounting for needed flexibility. It is not intended to be bureaucracy but development that keeps system integrity.

One retail company implemented SalesForce without instituting governance procedures. Any department could submit customization requests through a variety of means—sending email to IT staff directly, logging tickets, or catching developers in the hallway. No formal review of changes or impact assessment process existed.

In the space of a year, their environment had more than 300 custom fields, 75 custom tables, and many dozens of workflows built by various developers with inconsistent naming conventions. If a security update was necessary, they couldn't deploy it without breaking many customizations. Remediation in this emergency cost $870,000 and pushed off critical business projects by nearly a year.

"We had no inventory of customizations, record of why changes were made, or understanding of how to test," their CIO explained. "Some customizations did the same thing but differently for different departments. Others canceled each other out or had circular dependencies. What was meant to be one system had fragmented into dozens of solutions barely held together."

An effective governance model would have established well-defined, well-documented change request channels, architectural guidelines, impact analysis procedures, and a configuration management log for tracking all the customizations.

Training

The COE should develop and institute training programs that build system knowledge across the organization.

Effective training programs include:

- New user onboarding based on role
- Role-specific training paths
- Advanced capabilities training for power users
- Process-oriented training that connects system functions to business outcomes
- Continuous education on new features and capabilities
- Technical and administrative training for IT staff
- Training for trainers to scale knowledge

A manufacturing firm invested $3.2 million in a SAP S/4HANA implementation. Their training approach was building a library of videos and documentation within their intranet under the slogan "self-service learning for the modern workforce." Classroom training was delivered once during go-live, but no continuing program was put in place.

After launch, productivity was 32% lower after eight months on their new system. As the staff turned over, new staff were given partial knowledge transfer by other staff members who had themselves received only partial knowledge. New functionality was not utilized after the first significant upgrade.

The breaking point was reached when, during a compliance audit, auditors discovered that workers were keeping crucial manufacturing inventory information in spreadsheets due to their lack of knowledge about how to properly input it into the ERP system.

"We confused providing training with actually ensuring learning occurred," the HR Director conceded. "We had no way to validate comprehension, monitor progress, or even recognize knowledge gaps. We were technically in compliance with our training obligation, but completely ineffective at confirming real system knowledge."

The firm ultimately established an extensive training program with certification tracks, but not before it had spent $1.5 million on process remediation and compliance fines.

Support

The COE provides timely assistance as a part of an overall escalation path needed for system issues, questions, and enhancement requests.

Effective support includes:

- Documented triage paths and system problem resolution
- Answers to simple user questions and process guidance
- Handling improvement requests
- Troubleshooting data quality issues
- Coordination with vendors and development partners for complex issues
- System performance and health monitoring
- Usage tracking and adoption metrics
- Actively tackling impending problems when they are uncovered

For instance, if an organization sees support as simply repairing things when they are broken, then they are not leveraging the opportunities to see total system utilization. If they look at each support transaction as an opportunity to educate users and to spot training deficiencies or areas for system enhancement, then they will be creating value on a consistent basis instead of just problem-solving.

A financial services firm established a support team for their Microsoft Dynamics 365™ customer engagement implementation with solely technical administrators as staffing. Requests for support were evaluated on technical merit alone. Thus, if the system was functioning according to specs, problems were closed as "working as designed" regardless of whether or not they solved business problems.

The crisis happened in the middle of a large client onboarding project. Customer service reps were unable to set up the system to run a new financial product's approval workflow. The help desk kept closing their tickets, saying, "The approval system is working fine." What the technical team failed to understand was that the business required instruction on how to set up the workflow, not repair a damaged feature. This type of issue is where support escalating to the COE is a critical path.

The bypass introduced manual workarounds that caused compliance issues, slowed processing time, and ultimately the loss of a $20 million client.

"Our technical support people were troubleshooting technical problems and not business problems," the COO explained. "They could probably tell you whether a button was working correctly, but not whether pushing that button would give you the outcome you wanted. They were technologically correct but operationally incompetent."

After this costly failure, they reengineered their support model to incorporate the COE who were knowledgeable about both system capabilities and business processes, transforming support from reactive technical fixing to true business enabling.

Innovation

Business application systems are on a fast track for continuous development these days. The COE should continuously search for and adopt new system functionality that creates business value.

Innovation activities include:

- Tracking vendor software roadmaps and release notes for applicable new functions
- Researching industry best practices in system utilization
- Staying involved with user groups and relevant educational conferences
- Identifying and tracking of opportunities for improving system capabilities

- Piloting new features with specific user groups
- Creating business cases and value justifications for major improvements
- Coordinating the development of and implementing approved changes
- Measuring and reporting innovation results

For example, if a manufacturing company focuses their innovation on deploying new cutting-edge software, they might be overlooking an opportunity to get more value out of the software they already own. If, instead, they maintain a capability catalog that maps system capabilities to business processes, they can seek opportunities to utilize under-leveraged software functionality before evaluating new systems.

With these key activities, the COE helps your business system evolve from a static application into a dynamic platform that grows with your business on an ongoing basis.

A healthcare organization implemented Microsoft Dynamics 365™ for customer engagement but lacked any process for assessing and adopting new features. They paid their annual software maintenance fees religiously to the tune of approximately $400,000 annually, which entitled them to all new releases and functionality. But they never implemented any new releases.

Three years after go-live, they were found to be utilizing the same essential functionality they had at go-live, even though Microsoft had introduced dozens of major enhancements. They paid more than $1.2 million for improvements they never realized. Worse yet, the customizations that were put in place began to have serious issues due to the API and platform changes that were never modified during that three-year period.

This situation culminated when their competition began to provide a patient self-scheduling portal and mobile apps for schedule monitoring, new apps that were in their system for more than a year but never turned on.

"We used to treat our CRM like a static application, install once and simply keep it running," the IT Director said. "Nobody was keeping track of what was in every release, determining its business value, or encouraging use of the new functionality. We were paying for innovation we didn't utilize while we watched competitors leapfrog us using the very same software."

The client ultimately got so far removed from the initial implementation that they spent just as much money updating their system to account for new features and changes as they did on the initial rollout.

Assembling the Right Team

After you have determined the core disciplines of your center of excellence, the most crucial decision is the people who will make it a reality. No matter how well your COE organization and processes are designed, it's the talent who will determine its failure or success.

Good COEs need three types of participants, representing technology, business unit, and leadership, each offering key perspectives and competencies.

Technology Participants	Business Unit Participants	Leadership Participants
Solution architects: Design system architecture and integration approaches	Process owners: Understand end-to-end business processes across departments	Executive sponsors: Provide strategic direction and secure necessary resources
Developers: Customize and extend system capabilities	Subject matter experts: Provide deep knowledge of specific functional areas	Business unit leaders: Ensure alignment with departmental objectives
Data analysts: Ensure data quality and create reports	Power users: Demonstrate advanced system proficiency	IT leaders: Facilitate technical resources and integration

Technical support: Troubleshoot complex system issues	Change champions: Influence peers and drive adoption	Process improvement leaders: Drive operational excellence initiatives
Security experts: Maintain system integrity and compliance	Operational managers: Understand day-to-day business needs	Change management leaders: Facilitate organizational adoption

Key Roles Within the COE

Role	Responsibilities
COE leader	<ul><li>Develops COE vision and strategic plans</li><li>Secures and manages resources</li><li>Establishes governance frameworks</li><li>Measures and communicates COE performance</li><li>Creates training plans for team members</li><li>Coordinates with executive leadership</li></ul>
Business process experts	<ul><li>Analyze business processes supported by the system</li><li>Identify process improvement opportunities</li><li>Translate business needs into functional requirements</li><li>Validate system changes against business processes</li><li>Train users on process-based system usage</li><li>Develop process-based performance indicators</li></ul>
Trainers and change managers	<ul><li>Create comprehensive training materials</li><li>Deliver formal training to user groups</li><li>Develop self-service learning resources</li><li>Evaluate training effectiveness</li><li>Identify and address resistance to change</li></ul>

	• Communicate system changes and benefits • Celebrate and publicize success stories
Technical specialists	• Manage system security and configuration • Develop custom extensions and integrations • Create and maintain reporting solutions • Integrate with other enterprise systems • Troubleshoot complex technical issues

This balanced team structure creates a COE capable of addressing the full spectrum of system needs, from strategy to technical implementation, from process redesign to user adoption.

Measuring Training Effectiveness

Though most companies track basic training metrics like attendance and completion rates, meaningful measurement connects training explicitly to user competency and business outcomes. A holistic strategy for training makes sure your system education is all about capability building instead of an HR compliance activity.

Individuals learn in different ways, and acknowledging these variations significantly enhances training results:

Visual learners understand best through diagrams, pictures, and demonstrations. For them, provide:

- System flowcharts showing end-to-end processes
- Video recordings of important tasks
- Annotated screen captures with highlighted navigation pathways
- Visual dashboards showing competency progression

Auditory learners learn by listening and talking. Assist them by providing:

- Lecture-based lessons teaching system concepts
- Group discussion on business processes

- Opportunities to explain procedures back to power users

Hands-on learners who would rather learn by doing.

- Sandboxes for on-the-spot training
- Step-by-step guided exercises based on real-world processes
- Role-playing activities based on actual business data and scenarios
- Real-time feedback from power users or trainers

A financial services organization included learning assessment tests at the start of their CRM training process. They found that their sales force was very hands-on, whereas their finance group leaned toward visual. By modifying training techniques to these styles, they increased knowledge retention across the organization.

One-size-fits-all standard training is bound to fail since it does not account for different job roles, responsibilities, and skill levels. Successful training programs adopt personalization based on individual job roles. Specialized sessions may include:

- Executive dashboards and leadership analytics training
- In-depth workflow insight for process owners
- Comprehensive data guidelines for operations
- Administrator setup and configuration

Also offering flexible pacing as a part of the training supports varying learning speeds:

- Self-study modules for independent learners
- Organized cohorts or groups for team-based positions.
- Fast tracks to more advanced power user materials for technology-literate users
- Extended office hours for individuals requiring additional assistance

As a part of a system rollout, one manufacturing company charted 27 different learning paths throughout their enterprise, each with competency demands tied directly to job tasks. This created a targeted framework for how to develop and rollout training plans and also created a blueprint for new employee onboarding based on structured learning pathways.

Using assessments before and after training provides hard evidence of skill development and the effectiveness of the learning. Too few organizations offer these kinds of assessments at first, only to wish later that they had worked on accurately measuring true performance.

Effective assessment approaches include:

- Knowledge checks that test understanding of key software and process concepts
- Scenario-based assessments that evaluate application of knowledge
- Hands-on exercises that demonstrate practical system skills
- Role-specific competency evaluations
- Self-assessments that gauge user confidence and perceived competence of various areas

This scenario-based testing helps to understand the application of knowledge. The practical exercises demonstrate hands-on system skills. The self-assessments help evaluate user confidence and their own perceived competence.

For instance, if a health care organization is using simple multiple-choice exams to evaluate training success, then good scores will not equate to successful system use. If they utilize scenario testing, in which users have to show how they would use the system in real-life scenarios, then they'll know a lot more about true user competency.

By holistically addressing learning styles, individualizing training pathways, and utilizing stringent assessment techniques, organizations turn training from a requisite expense into an investment that will increase system adoption and the ability to become more self-sufficient in the future.

Ongoing learning programs must offer progression toward sophisticated system capability. One company took this approach to the next level by regularly implementing what they called "system challenges," fun assessments that challenged users on their knowledge of important system functionality. Not only did these lighthearted competitions reveal areas of weakness, but they also made learning a competitive activity that employees actually looked forward to. Challenges were then followed up with targeted five minute or less micro-learning sessions that plugged specific gaps, ensuring continuous skill development. This kind of layered approach avoids overload and allows company-wide skill levels to increase over time.

When one company rolled out its new ERP system, they used the typical training playbook: classroom training, user guides, and a support hotline. Six months down the road, however, they were confronting issues of adoption. Despite spending a great deal of money on training, their employees were only utilizing core functions that were used multiple times per day and shying away from less used workflows and advanced features that had the potential to deliver more business value.

The moment arrived when their training manager began to employ games within the organization. "People enjoy getting to levels, earning badges, and being recognized," she explained. "What if we used the same principles for our system training?"

The outcome was an extensive internal certification process that revolutionized their system expertise approach. Instead of training being something that employees "finished," the certification program defined a clear progression path with three levels of expertise: foundational, advanced, and expert. The demonstration of specific skills had to be shown at each level through both knowledge assessments and real-world applications.

What rendered the program effective was not the idea; it was how that company integrated it into their organizational culture and career advancement pathways.

- Employees who achieved advanced or expert certifications were granted higher system privileges, which allowed them to perform more advanced tasks independently of IT.
- Certification success was applauded in leadership meetings and corporate newsletters, providing public recognition to the employees for their success.
- Most significant, certification levels were mentioned in job postings and taken into account when deciding about promotions and pay increases. This established concrete career benefits for acquiring system knowledge.

"We had to make it meaningful," the training manager said. "If certification is just a piece of paper, nobody cares. But when it's tied to real opportunity and even monetary reward, then it's something that people desire."

The effect went well beyond better system usage. The program established a defined career progression for workers who wanted to expand their technical acumen, enhancing retention of key talent. It also allowed the company to develop potential internal prospects for its center of excellence. The IT department hired from the pool of expert certified individuals, offering an even greater career trajectory for some employees.

Internal certification programs work because they make system training a process of professional growth. They define precise requirements for measuring competency, create a common language for levels of skill, and provide tangible recognition. For companies looking into this strategy, the trick is to make sure certification standards are based on actual practice, rather than textbook knowledge.

One manufacturing worker explained the difference: "In our previous training, we'd sit through someone going through a process and then take a multiple-choice test. Now, to become certified, I have to actually do the task and explain to you why every step is important to our business process. It's more difficult, but I actually remember what I learn."

The best certification programs change over time, incorporating new skills as system functionality changes and business requirements shift.

Empowering Business Users to Take Ownership

The shift from dependency to autonomy is not an arbitrary process. It requires a conscious plan with several essential elements.

To begin with, knowledge transfer requirements must be made contractually clear in consultant contracts. That is, documentation norms, training expectations, and knowledge-transfer sessions must be contractual necessities and not discretionary add-ons. One retail firm has a clause where 15% of the consultant fee is withheld until knowledge transfer is assessed by testing the practical competencies of the internal team.

Second, internal staff must be paired with consultants from day one, not just in the handoff stage. These pairings create natural knowledge transfer situations and prevent consultants from becoming isolated knowledge silos. One manufacturing firm takes it a step further by assigning each internal team member working with a consultant specific knowledge acquisition objectives.

Third, retain all intellectual property developed throughout the engagement. This encompasses documentation, bespoke code, configuration settings, and implementation tools. One health provider learned this lesson the hard way when their consultant left with proprietary configuration tools, and they had to rebuild these assets from scratch.

Finally, the consultant relationship should reach maturity rather than coming to an end prematurely. As internal capability grows, consultants can transition their role from hands-on implementation to strategic advisory roles, providing occasional checks and guidance on difficult issues while internal personnel handle day-to-day operations.

This incremental handover of authority is a model that is sustainable, as organizations maintain contacts with beneficial consultant allies while building up the internal know-how towards autonomy.

Creating a Thriving Center of Excellence

Developing in-house competency is not only a matter of personal knowledge—it's a matter of establishing organizational systems that foster ongoing learning and development. That's where a COE can help.

However, most organizations fail to sustain momentum after the initial excitement of creating a COE has worn off. Members get drawn back into daily operations, knowledge sharing occurs haphazardly, and the COE slowly becomes obsolete. The secret to avoiding this is to establish meaningful incentives that encourage people to continue contributing to the COE over the long term.

Here is an example of a company that had this issue. Six months following the go-live of their Dynamics 365 COE, adoption was slowing down. Volunteering subject matter experts were bogged down in their day-to-day activities, and knowledge-sharing sessions were poorly attended.

"We knew we were requesting individuals to contribute a great deal of effort and time without formal recognition," their Director of COE clarified. "It was being handled as an extra on top of their full workloads."

The turnaround started when they launched a multilevel incentive approach that tackled extrinsic as well as intrinsic motivation. They established an official "CRM Champion" title with three levels, each offering progressive recognition and reward. Champions received branded items, were featured in company announcements, and enjoyed priority access to exclusive events and senior-level training.

Most importantly, they re-engineered jobs to allow protected time for COE work. Champions were officially allocated 10-20% of their time for system improvement work so that these responsibilities were not added on top of their other workloads. This time allocation was built into performance plans and was reinforced by managers.

Yet the most powerful motivator was linking COE participation to career growth. System knowledge and contribution to the COE were made part of promotion criteria, and COE positions were reframed as developmental assignments with exposure to senior leadership.

"When people realized that COE membership would mean career advancement, it was a whole new ballgame," stated the Director. "We went from begging for attendance to having a waiting list of folks who wished to be members."

Another company tried another strategy, trying to develop a unique community identity for their COE members. They developed quarterly innovation days when COE members were allowed to present their work to the larger organization. These were much-anticipated events, which provided COE contributors with an opportunity to exhibit their expertise and innovation.

This example illustrates a fundamental fact—rewards and recognition are worthwhile, but the most compelling incentives are those connected to career progression and professional growth. When COE involvement furthers careers, it becomes a privilege rather than an obligation.

Effective incentive schemes must incorporate rewards that recognize contribution and provide visibility for COE members. These may be as simple as acknowledgment in company newsletters or as sophisticated as award programs that reward excellence.

Career development programs that enable members to acquire valuable skills are very effective. Specialized training, certification programs, conference attendance, and collaboration with vendor product groups are all compelling incentives for employees who will simultaneously learn additional skills.

Career development issues that formally link COE contributions to promotional prospects incentivize team members. These inducements may include the addition of COE responsibilities to job descriptions or the creation of career paths specifically for system specialists.

The key is to align incentives with both organizational goals and individual motivations. As one COE leader put it, "We had to provide an answer to the question 'What's in it for me?' from every contributor's perspective. When we found those answers, engagement escalated."

Establishing a Governance Framework for Continuous Improvement

An effective center of excellence requires more than enthusiastic participants—it requires formal governance that allows for repeatability. Without governance, even the most enthusiastic COE will lose direction and become a hindrance rather than an enabler.

Let's look at an example of a company that found out the hard way. Their first SAP COE was very technically focused on governance—change control, release management, and system stability. Their burdensome change approval processes meant nothing got broken during releases, but it also became a bottleneck for valuable business improvements. Things got stuck in approval cycles.

"We were so focused on safeguarding the system that we forgot why we needed it in the first place, to create business value," their Director of IT reflected. "Our governance was just risk aversion, disguised as a process, not value creation."

The wake-up call came when one of the divisional VPs bypassed the COE entirely, hiring an outside consulting firm to create a reporting solution because the formal enhancement process was moving too slowly. They needed data visibility to operate. The shadow system, after being implemented, also created data inconsistencies and security risks, precisely the kind of problems that governance was designed to prevent.

The organization changed its direction by consolidating and even eliminating some of the governance steps in order to put equal emphasis on getting new features delivered at speed and technical quality. They adopted a balanced approach with a set of requisite elements.

First, they established clear decision rights using a RACI model (responsible, accountable, consulted, informed) for different types of system changes. They eliminated the confusion that had led to bottlenecks in the past, ensuring the right people made decisions at the right level.

Second, they established service levels and performance metrics for the COE itself and held it accountable for response time, time to approval, and release schedule. Those metrics were examined every month by a steering committee to grade the performance of the COE.

Third, they scheduled regular surveys in which business stakeholders and technical teams provided feedback on whether the COE was achieving its objectives. These open, honest discussions between business requirements and technical systems owners allowed issues to continually surface, which prompted course corrections.

"The change in governance was much needed," reported the COE Leader. "We shifted away from being viewed as system police to being valued business partners. And ironically, we actually improved system stability because business units no longer felt like they needed to develop shadow systems or workarounds." Good governance is all about balancing enablement with providing consistency—without stifling innovation.

As one executive said, "Governance sounds political, but really it's about enabling innovation to happen. When we have good guardrails in place, people feel more secure to take risks and try something new."

Designing Effective Training Programs

Training is the bridge between business value and system capability, but only when it is designed to create real expertise, not shallow familiarity. Successful training does much more than include presentations or reading materials; it should construct deep capabilities with a direct impact on business readiness and continuous improvement.

Imagine a logistics company whose warehouse management system training is initially comprised of mostly instructor-led demonstrations followed by user guides provided afterward. Students would leave training understanding how to duplicate steps in the system, but would freeze when confronted with real-life situations or exceptions on the warehouse floor.

"We realized we had been training people on the developed system instead of on how to use the technology," their Director of Operations said. "It was like trying to learn how to swim from videos and textbooks without ever getting in the water."

The firm redesigned its approach by implementing a structured, three-step training process that built up expertise step by step.

The foundation phase was first, and provided an equal ground of hands-on, lab-based system familiarization among all the users. This included system navigation, key functionality, organization-specific configurations, data relationships, security practices, and support procedures. This was the groundwork that got everyone speaking a common language and mastering the fundamentals before moving to role-based training.

The role-based phase layered on scenario-based hands-on system processes and functions that are applicable to specific job functions. Warehouse personnel learned picking and packing processes, customer service personnel learned return processing, and accounting personnel learned invoicing and reconciliation. The intensive focus allowed users to rapidly achieve competency in the activities most applicable to their day-to-day roles.

The third, advanced level, developed specialized expertise for power users, administrators, and COE members. They acquired system configuration, reporting, workflow design, integration, performance, and advanced troubleshooting knowledge. This level developed a distributed network of technical team members to serve the wider user base.

The structured process provided users with the relevant skills needed to carry out job responsibilities and created career paths for users showing potential to move into more technical roles.

The most dramatic change was in how training was being presented. This company shifted from demo-based classes to experiential learning classes that engaged users directly with the system. They created sandbox environments where users could practice developing new features or editing workflows without affecting production data. They crafted guided exercises that built specific skills step by step. They created scenario-based learning exercises that involved using several system functions to solve real-world business problems.

"The difference was immediate," the Training Manager reported. "Rather than sitting and observing someone else configure the system, our employees were configuring it themselves—making errors in a controlled situation and learning how to recover from them. By the time they reached the warehouse floor, they'd already rehearsed many of the situations they'd face."

The best training labs and exercises cover three fundamental skill sets: technical, process, and soft skills.

Technical skills should be aligned with the mechanics of effective use of the system, navigation, use of features, configuration, reporting, data management, integration, and troubleshooting. These skills allow users to enhance and diagnose the system appropriately and effectively.

Process skills relate system functionality to business processes. They establish good comprehension of how particular system activities play a part in end-to-end business processes. Process skills make the users think beyond single tasks to realize how they affect upstream and downstream tasks.

Soft skills support efficient system utilization and teamwork. These skills include technical and business department communication, problem identification and troubleshooting, requirement gathering and documentation, stakeholder management, change management, knowledge transfer, and continuous learning.

By addressing all three skill categories, this company produced well-rounded system specialists who were not only able to run the system but also utilize it to make significant future business improvements in a continuous development model.

As their Training Manager commented, "We eventually came to understand that familiarity with the system is not the same as being capable of running it effectively. Our new approach generated real capability, not just familiarity."

Upskilling and Reskilling: Transforming Employees into System Experts

Not everyone will be system experts, nor does everyone have the aptitude or interest. The secret to effective upskilling is to select the appropriate talent—those who will give you the greatest return on your investment in their training.

Companies often select candidates for more advanced system training based on technical aptitude, choosing those who have experience with technology. While these individuals learn system mechanics with ease, something is missing.

"They might be able to do the technical work just fine, but they couldn't tell us why the steps mattered in our business," the Training Director remembered. "If other users posed questions like, 'How does this enhance client experience?' or 'Why do we have to input this information?' our technologically-adept experts didn't have those answers."

The internal talent shifts began when the company changed the selection criteria to weigh business acumen just as heavily as technical ability. They started searching for applicants with a keen insight into workflows, regulatory mandates, and organizational procedures—individuals capable of relating system features to actual business requirements.

"The difference was night and day," the Director said. "When our internal team understood why system functionality was needed, adoption skyrocketed. They were able to describe features in terms of business value instead of just technical specifications."

Organizations must look beyond technical capability in shortlisting candidates for upskilling. Top system experts possess a mix of technical knowledge and business understanding, application skills and communication skills.

Analytical thinking is especially important as it allows the individual to identify patterns, diagnose complex problems, and understand how the various components of a system interact. Candidates who are able to trace problems back to their origin and recognize relationships among problems are most likely to make great system experts. It's more about the way their brain is wired to solve puzzles than it is about their ability to write code.

Communication skills should not be ignored, since system experts often have to act as translators between the technical and business worlds. They should be able to describe technical items in business language and interpret business requirements into technical specifications. This translating skill is usually what distinguishes highly successful individuals from those who know the system yet are not as successful in gaining adoption.

One healthcare organization employs a unique method of selecting candidates for upskilling. They search for people who are well respected by their colleagues and often consulted for assistance—individuals who are already operating as unofficial knowledge hubs. These natural helpers, the company found, were more likely to possess the credibility and communications abilities to pass on knowledge effectively. Taking those individuals and investing in their technical skills allows for development of highly skilled team members who are able to function long-term within an organization.

Growth mindset is one other trait to seek out. Individuals who consider challenges, something they may have never done before, as a chance to learn instead of a mental and emotional hurdle too big to take on will succeed in a position that involves continually adapting to new features and functions. This is the reason why individuals with sound business acumen have a head start in acquiring system proficiency, no matter their technical skills.

Consider what happened when a national financial services firm discovered the principle through comparison. When they rolled out a loan management system, they trained two groups of users: senior loan writers with little technical expertise, and technologically savvy IT personnel with little loan writing expertise. While the IT staff learned to navigate the system more quickly at first, the senior loan officers learned to use the system to produce the next phase of real-world business scenarios better than the members of the IT team.

"The technical people could tell us what buttons to push, but the loan officers knew why we needed to require certain data and how to better change the requirements to make it more effective," the Training Manager clarified. "They saw right away how system functionality mapped to procedures they'd been doing for years." The context made the technical education more concrete. Once workers can directly correlate new system functionality into known business processes, add-on learning becomes more natural and applicable.

The best training plans take advantage of this relationship by using training that starts with business processes, not system functionality. Rather than addressing how to create a customer record in the system, effective training is positioned within the context of business activity, for example, how to onboard a new customer.

Using real business situations as hands-on labs is an effective approach to training as it simulates real-world usage instead of theoretical practice. Many times, these role-based training tracks make sure that employees learn what is applicable to their immediate job first before expanding to more general skills.

Great training always covers the "why" of system operation, explaining the business purpose each feature is intended to achieve. Context elevates system features from mere procedure to valued assets for business success.

When developing training, outlining the business value of every system's function should be required, as doing so adds much greater value than merely instructing on how to input data. By leveraging proven business experience as the foundation for system knowledge, firms accelerate learning and ensure system capability is translated into business value.

Harnessing the Power of Troublemakers

They can be found in every organization—the power users who naturally grasp new systems and immediately begin to push the boundaries. They try out undocumented features, develop novel solutions to business problems, and become the undesignated authority when others need help. Power users are a valuable resource that can dramatically advance system adoption and innovation if companies identify and leverage them correctly. We affectionately call these individuals the troublemakers because they are never content with the system or processes as they are, instead constantly looking for ways, and encouraging others, to push boundaries even when they may not be assigned to that type of role internally.

What distinguishes these troublemakers from the rest are their unique traits.

They are explorers who don't simply wait to be formally trained or instructed. They actively experiment with system behavior, discover hidden aspects, and devise innovative solutions for solving business issues. At a global financial services firm, one of the champions discovered and documented more than 30 helpful features or shortcuts not included in official training. Another user created a keyboard shortcut card to attach to the screens of their colleagues for quick access to frequently used functions.

They adapt easily to change, rapidly incorporate new skills into their work routines, and are frustrated when others do not do the same or with governance that may be viewed as a blocker to their innovation. When a bank launched a key platform upgrade, these champions were in a position to learn the changes and lead their colleagues through the process, minimizing productivity loss and shortening the time to value.

They are data-driven decision-makers, regularly examining trends, troubleshooting, and gauging results. One power user at a company created a daily dashboard that allowed customer service representatives to prioritize client contact based on recent history, which boosted customer satisfaction ratings considerably.

However, not all power users expend their energy constructively. When they do not feel enabled, they are also those who create shadow systems, rogue workflows, their own Excel spreadsheets, and sometimes disastrous workarounds that circumvent official processes. Though such activity may be frowned upon, it usually indicates legitimate needs that have gone unaddressed.

This scenario manifests in a process sometimes referred to as vibe coding, where these creative individuals instinctively hack together solutions, guided more by intuition and firsthand experience than by formal processes or documentation. These so-called troublemakers, once on the periphery, now find themselves uniquely positioned in this era of AI-driven tools. Artificial intelligence has amplified their capabilities, giving them new abilities to prototype, automate, and share solutions at scale. Instead of cobbling together shadow systems in isolation, they can leverage AI to rapidly translate workflow into enhancements or new features, often outpacing official development cycles. Organizations that recognize and harness this vibe coding energy can benefit enormously, turning what was once seen as troublemaker behavior into practical innovation that can bridge the gap between business and technical execution.

To put this in perspective, imagine one company that discovered this when they were exploring why their R&D department had constructed their own issues-tracking system when they had access to the enterprise project management platform. The immediate reaction from the IT department was predictably defensive; the shadow system violated data governance policies, brought in compliance risks, and duplicated information that was already available in the official system.

"Our first impulse was to shut it down and force everyone to revert to the standard system," the IT Director stated. "But we stepped back and asked ourselves why they invented this parallel process to begin with."

Through observation and non-judgmental interviews of R&D personnel, they learned that the department had valid requirements that were unmet. The enterprise system didn't have the specialized fields required to capture regulatory submission information, lacked the data visibility feature that researchers required, and was encumbered by inefficient workflows from other departments that decreased their productivity.

Instead of merely outlawing the shadow system, this company converted these troublemakers into useful agents within the COE. They asked the creator of the shadow system to head a process improvement team tasked with optimizing the module within the enterprise system.

"It was a total mindset shift," the R&D Director described. "Rather than being flagged as non-compliant, our staff were suddenly praised for having discovered essential functionality gaps. The same innovative energy that had gone into creating workarounds was channeled into enhancing the official system."

The company initiated a number of measures to capture this innovation.

They held interviews to discover pain points, observed unofficial workflows to determine system gaps, and analyzed the homegrown solutions to learn about unmet needs. This digging unearthed real problems that had to be solved.

Next, they integrated these innovative troublemakers into system improvement conversations, directing their creativity into approved channels. Their in-depth understanding of day-to-day problems paired with their creative problem-solving skills resulted in improvements that were of benefit to the whole organization.

This approach not only improved system functionality, it also transformed organizational culture. Troublemakers were converted into contributors.

"We realized that these shadow systems were expressions of unmet needs, rather than acts of rebellion," the IT Director said. "By meeting the unmet needs rather than merely prohibiting the workarounds, we strengthened our organization, encouraged innovation from employees, and built tighter relationships within our business units."

In my experience, workers who focus on creating unofficial workflows, or workarounds, and their own tools are generally your most engaged and creative employees. Most times, they are acting out of genuine business needs, or an ignorance of what is currently possible with the tools at their disposal, not out of willful insubordination. By recognizing these needs, organizations can harvest great benefits by aligning these creative efforts with official flows.

Training on Communication to Bridge the Technical and Functional Gap

Enterprise systems sit at the intersection of business and technology. Communicating across that gap requires special skills. Technical teams typically don't have the skill to describe system capabilities in business terms, and business users often don't have the technical skill to articulate their needs in terms that can be implemented by technical teams. This gap in communication is more often than not the root cause of mismatched expectations, inadequate solutions, and failed implementations.

One Microsoft consulting firm encountered this problem with its CRM implementation. Even with a seasoned technical team and business-aware users, projects would invariably fail due to misunderstandings and expectation gaps. Business stakeholders signed off on requirements documents that then generated solutions that never addressed their true needs. Technical teams implemented precisely what was asked for, only to be told, "That's not what we wanted."

"We knew we were speaking different languages," stated their Director of IT. "Our technical team discussed the features and capabilities of the system, and our business users were speaking in terms of outcomes and process. Both thought they were being clear, but in reality, they were talking past one another."

They changed their strategy by performing large-scale communication training that focused on three essential skills.

They started by teaching business users how to write clear requirements with an emphasis on technical outcomes, rather than just software features. Instead of requesting "a dashboard with KPIs," users were taught to request "a visual way for sales managers to view underperforming lines of business based on revenue and identify the root causes by seeing detailed data for transactions." This outcomes-based approach gave technical teams the context they needed to create correct solutions. They performed exercises together to write requirements based on real-world and fictional company needs.

This training covered essential requirements skills like distinguishing needs from features, stating business outcomes correctly, prioritizing and testing requirements, documenting dependencies, and validating requirements with stakeholders. In one workshop exercise, attendees rephrased vague requirements like "user-friendly interface" into clear, measurable specifications.

Second, they trained staff to ask the right questions in problem-solving conversations. This involved separating symptoms from root causes, utilizing open-ended questions to seek context, recognizing assumptions, and validating the understanding before moving forward. When users asked for new features, they were taken through a systematic conversation by those who had been trained to do so in order to determine the genuine business requirement within the request.

This change in the way requirements were gathered often revealed that existing system functionality was capable of meeting the needs with minor adjustments, averting unnecessary development. One department requested a custom reporting module only to discover through careful questioning that a standard report with different filtering would provide the exact information that they needed.

Third, they developed strategies for documenting technical and business terms. Technical teams were coached to explain concepts in business terms, while business users were introduced to key technical concepts that affected the way they wrote requirements. Both groups practiced, in fun and many times comedic ways, tailoring their communications to different audiences to enhance understanding.

"The moment of truth was when we broke out of looking at this as a technical problem and understood that it was a communication gap," explained the Director. "We needed to establish bridges among the various worlds."

They also established formal feedback loops and retraining plans that provided alignment between business requirements and technical solutions. Formal user visualization sessions enabled business users to see solutions under development and offer feedback, and post-implementation meetings brought out lessons learned to enhance future communications. As business and technical teams reached a common vocabulary and understanding, organizational cooperation improved. Technical teams more effectively grasped business processes, while business users formed more realistic system capability expectations.

"We've established a virtuous circle," explained the Director. "Greater communication creates better solutions, which reinforces trust in the team and generates even greater open communication. Something that began as a project management tool has taken on a life of its own as a profound, deep-seated cultural change in the way our technical and business teams collaborate."

Conclusion

Enterprise systems are never truly complete. They need to constantly change to address ever-evolving business requirements, incorporate technology updates, and provide more value over time. The secret to maintaining this growth is the development of strong feedback loops that invite open and honest user feedback and translate it into tangible system changes.

One manufacturing company that invested millions in its ERP implementation deemed the project complete when it finally went live with the system. The implementation team stopped meeting, outside consultants left, and the focus shifted to newer projects. The system became a static entity, kept on life support by the technology department but not built upon.

Users created more and more elaborate workarounds over time to address unmet needs. Shadow systems spread. Adoption decreased as newer employees questioned practices that were no longer relevant to business realities. A once cutting-edge system slowly evolved into a hindrance to the business's ability to grow.

A breaking point was reached when a new CIO came in and saw these symptoms and introduced a holistic COE to power ongoing improvement. "We had to shift our mindset from believing the ERP project was complete to understanding that our ERP journey continues," she said. "The system had to evolve with our company, and that meant new ways to glean feedback, embrace it, and turn it into improvements."

They established quarterly system roundtable sessions in which users from throughout all departments gathered with the center of excellence team to discuss experiences and needs. Group sessions revealed trends and priorities not evident from simple improvement requests. Cross-department conversation exposed participants to other departments' points of view, so solutions were more comprehensive and focused on end-to-end processes instead of stand-alone functions.

They set up innovation labs where power users could learn and test new capabilities in sandbox environments. One production planner used the opportunity to develop a new scheduling methodology that ultimately saved time when implemented in the production system.

They built feedback loops right into the system itself through lightweight forms that allowed users to report an issue, suggest an improvement, or share a success story without ever leaving their workflow.

To weigh innovation alongside governance, they established structured review processes for suggested enhancements. All suggestions were rated on business value, ease of implementation, alignment with strategic goals, and possible risks.

The COE also served as a continuous improvement team. They hosted monthly knowledge exchange forums where users shared techniques, workarounds, and solutions. Recordings of the sessions created a growing set of real-world documentation, while shared workarounds revealed system gaps that needed to be addressed.

"The magic happens when you connect user feedback directly to system evolution," observed the CIO. "When people see that their suggestions actually get implemented and can experience the resulting operational improvements, they become more invested in the continuous improvement process. It becomes self-sustaining."

As we've explored throughout this chapter, the journey from implementation to realizing sustained value is fundamentally about training and changing the perception of people, not just about technology. Developing internal expertise through a well-structured center of excellence, strategic upskilling for the troublemakers within your organization, coupled with a culture of continuous learning will transform your enterprise system from merely an application into a platform that empowers and evolves with your business.

The key principles we've covered include:

- Internal system ownership is essential for long-term success.
- A center of excellence (COE) for business applications provides an organizational structure for system governance, knowledge management, user support, and continuous improvement.
- Strategic upskilling and reskilling create a workforce that can fully leverage and continue to improve on your enterprise system.
- Power users can drive adoption and innovation when properly identified and supported.
- Addressing troublemakers and shadow systems requires understanding underlying system gaps and needs rather than simply forced compliance.
- A culture of continuous education ensures your organization builds on its successes rather than starting over with each system change.

Organizations that successfully apply these principles shift from a cycle of third-party dependency and disappointment to one of strategic advantage. They reduce costs while simultaneously increasing organizational agility, enabling them to respond faster to changing business needs and market conditions.

The most valuable outcome isn't just financial; it's the transformation of your enterprise system from a technical burden into a strategic asset that creates a sustainable competitive advantage. Furthermore, by investing in your people you ensure that your technology investment continues to deliver value long after the initial implementation.

Reflection Questions

1. How dependent is your organization on external consultants for system knowledge, routine system support, and enhancement?
2. What current mechanisms exist for investing and training your in-house system expertise?
3. Who are the natural power users in your organization, and how are you leveraging their capabilities?
4. How effectively does your training approach balance technical, process, and soft skills?
5. What shadow systems or workarounds exist in your organization, and what unmet needs do they reveal?
6. How would you structure a center of excellence to meet your organization's specific needs?
7. What incentives could you implement to encourage ongoing skill development and knowledge sharing?
8. Who are the troublemakers in your company and how can you transform them into valuable contributors to system enhancement?
9. What feedback loops do you have or need to establish in order to ensure continuous system improvement?

Conclusion

You've made it to the final pages, perhaps feeling a little battered by the cautionary tales outlined in this book. If this felt at times more like an intervention than a gentle how-to guide—good. This book was intended to be a rescue mission.

We have dissected the raw truths. We have asked hard questions. We've challenged you to strip away illusions and face what's lurking beneath the surface.

The message is simple but urgent. Transformation doesn't happen because you bought new business software or drafted a glossy new strategy document. Transformation is gritty, daily, relentless, and only fueled by a willingness to confront dysfunction, to uncover the exceptions, and to invest, over and over again, in the messy business of people.

At the start of this book, I presented a jarring reality—companies globally will squander more than $2 trillion on unsuccessful ERP, CRM, and other digital transformations annually. In these chapters, through real-life cautionary tales, we've lifted the lid to find out why so many initiatives fail in spite of good intentions and expensive investments.

The solution, we have seen time and time again, is not in the technology itself but in the human dynamics that accompany it. Dysfunctional processes, mismatched expectations, organizational resistance, and negative leadership dynamics cannot be solved by even the most modern software. Technology magnifies what is already present in your organization, the good and the bad.

In that conference room I mentioned in our first chapter, where managers were confronted with the bitter reality of a failed implementation, the source issue wasn't a technical one. It was the people. And this cycle is repeated in conference rooms across the globe on a daily basis.

But it doesn't have to be this way.

The most successful projects I've seen in decades of consulting have adhered to the precepts we've discussed here. They've kept the objectives clear and business-strategy focused. They've offset visionary leadership with execution discipline. They've staffed on capability, not title. They've understood that requirements are not technical blueprints but the language of business value. And above all, they've invested as much energy in their people and processes as their technology.

As you consider your own transforming experience, ask yourself which of these seemed the most like a mirror into your organization:

- Is your leadership team balanced with integrators and visionaries?
- Have you created a Project Bill of Rights that creates psychological safety?
- Do you have full traceability of your requirements?
- Are your requirements coming from quantifiable business results?
- Is your view of each requirement well written and aligned with reality?
- Can the vision of the engagement be articulated and shared by the entire organization?
- Are you building internal capacity through training and upskilling?
- Have you addressed the data complexity behind your integration strategy?

This book is the chance to change your launch sequence. If you're reading this, hopefully you're not just drowning; hopefully you're the catalyst behind a complete shift in the way you're engaging ERP, CRM, and digital transformation projects within your organization. The tools, the frameworks, the provocations, these are yours now. Use them to challenge your teams to get beyond comfort and change.

Keep in mind that digital transformation is not a point of arrival, but an ongoing process. It is not about software implementation but about changing the way your organization operates in all processes. That change occurs when you create the human foundation on which technology can thrive.

As you proceed, I want you to be as deliberate about developing your leaders, processes, and people as you are about developing software systems. Understand end-to-end requirements and traceability. Implement your center of excellence. Establish formal knowledge transfer processes. Establish internal competencies that reduce dependence on external experts. Most importantly, create a culture where the people are more valued than the software, where risk can be spoken of without fear of reprisal, and where learning from failure is viewed as progress.

By concentrating on the human factors (the leaders, teams, and cultures that implement change), you set your organization up for long-term competitive success in a constantly changing technology environment.

My hope is that this book is both a warning and a guide—a warning regarding the pitfalls that have sidetracked so many projects before yours and a guide to successfully navigate those challenges. The principles we've discussed have aided thousands of organizations in changing not only their technology but their operations, their cultures, and ultimately, their results.

Let the other organizations waste their trillions. You don't have to. You have seen behind the curtain. Now, build the disciplines, start the conversations, and make the hard incremental changes that will anchor your success—not as a one-time leap, but as a relentless rise.

This is your guide to launch.

Resources

American Foundation for the Blind. (2023, June 27). Happy birthday, Helen!. American Foundation for the Blind. https://www.afb.org/blog/entry/happy-birthday-helen

Arthur, C., & Baxter-Reynolds, M. (2011, October 14). BlackBerry outage: Faulty router suspected. The Guardian. Retrieved from https://www.theguardian.com/technology/2011/oct/14/blackberry-outage-faulty-router-suspected

BrainyQuote. (n.d.). Peter Drucker quotes. Retrieved [date you accessed it], from https://www.brainyquote.com/authors/peter-drucker-quotes

Breyley, D. (n.d.). *No more 'Guess Who' with GWC™*. EOS Worldwide. Retrieved from https://www.eosworldwide.com/blog/no-more-guess-who-gwc#:~:text=Capacity%20to%20do%20it%3F,this%20issue%20quickly%20and%20objectively.

Blaney, B. (2024, November 26). What is a 3-way match in accounts payable & why should you use it. Tipalti. https://tipalti.com/resources/learn/3-way-match/

Bojinov, I. (2023, November-December). Keep your AI projects on track. Harvard Business Review. https://hbr.org/2023/11/keep-your-ai-projects-on-track

Boston Consulting Group (BCG). (2020). Flipping the odds of digital transformation success: Leading in the new reality. https://web-assets.bcg.com/c7/20/907821344bbb8ade98cbe10fc2b8/bcg-flipping-the-odds-of-digital-transformation-success-oct-2020.pdf

Bloch, S.Blumberg, and J. Laartz, "Delivering large-scale IT projects on time, on budget, and on value," McKinsey on Business Technology, October 2012.

DeMarco, T. (n.d.). The process of writing down the requirements is the process of determining them.

Flyvbjerg, B., & Budzier, A. (2023, September). Why your IT project may be riskier than you think. Harvard Business Review. https://hbr.org/2011/09/why-your-it-project-may-be-riskier-than-you-think

Grove, A. S. (1995). High output management (Paperback ed.). Vintage Books.

Hayes, A. (2021, August 12). Blackberry addiction. Investopedia. Retrieved from https://www.investopedia.com/terms/b/blackberry-addiction.asp

Imtiaz, K. (2023, August 6). What is security posture? CrowdStrike. https://www.crowdstrike.com/en-us/cybersecurity-101/exposure-management/security-posture/

Intel Corporation. (n.d.). Our values. Intel. https://www.intel.com/content/www/us/en/corporate-responsibility/our-values.html

Janis, I. L. (1982). Groupthink: Psychological studies of policy decisions and fiascoes (2nd ed.). Houghton Mifflin.

Mathews, S. (2022, June 19). How Bezos thinks – All his frameworks and mental models. Leading Sapiens. https://www.leadingsapiens.com/bezos-on-failure-decision-making-life/#:~:text=Disagree%20and%20commit?%E2%80%9D%20By%20the,need%20to%20worry%20about%20that.

McKinsey & Company. (2018, October 29). Unlocking success in digital transformations. https://www.mckinsey.com/capabilities/people-and-organizational-performance/our-insights/unlocking-success-in-digital-transformations

Minnaar, J. (2023, November 12). Musk's algorithm to cut bureaucracy. Corporate Rebels. https://www.corporate-rebels.com/blog/musks-algorithm-to-cut-bureaucracy

PwC. (2024, June 24). Global Workforce Hopes and Fears Survey 2024: Workers are ready for change. Are leaders ready to engage them? https://www.pwc.com/gx/en/issues/workforce/hopes-fears-survey-2024.html

Robinson, C. (2025, January 30). Helicopter, jet collision recalls 1982 Air Florida disaster that left 78 dead. WBAL. https://www.wbaltv.com/article/air-florida-disaster-potomac-washington/63615851

Saint-Exupéry, A. de. (n.d.). If you want to build a ship, don't drum up people to collect wood and don't assign them tasks and work... Goodreads. https://www.goodreads.com/quotes/384067-if-you-want-to-build-a-ship-don-t-drum-up

Seth, S. (2024, April 14). BlackBerry: A story of constant success and failure. Investopedia. Retrieved from https://www.investopedia.com/articles/investing/062315/blackberry-story-constant-success-failure.asp

Standish Group International. (2020). *The chaos report*. Copyright © 2020 The Standish Group International, Inc. All rights reserved.

Tunguz, T. (2017, November 12). Disagree and commit – A management principle for highly functioning teams. LinkedIn. https://www.linkedin.com/pulse/disagree-commit-management-principle-highly-teams-tomasz-tunguz/

Whitmore, A. (2023, November 28). Digital transformation waste bill expected to be $2 trillion by 2026. Business Matters. https://bmmagazine.co.uk/opinion/digital-transformation-waste-bill-expected-to-be-2-trillion-by-2026/

Wickman, G., & Winters, M. C. (2016). Rocket fuel: The one essential combination that will get you more of what you want from your business (Paperback ed.). BenBella Books.

www.ingramcontent.com/pod-product-compliance
Lightning Source LLC
Chambersburg PA
CBHW071547210326
41597CB00019B/3148